P9-DDV-045

BIODIVERSITY

A Reference Handbook

Other Titles in ABC-CLIO's
CONTEMPORARY WORLD ISSUES
Series

Abortion, Second Edition	Marie Costa
American Homelessness, Second Edition	Mary Ellen Hombs
America's International Trade	E. Willard Miller and Ruby M. Miller
Campaign and Election Reform	Glenn Utter and Ruth Ann Strickland
Children's Rights	Beverly C. Edmonds and William R. Fernekes
Crime in America	Jennifer L. Durham
Domestic Violence	Margi Laird McCue
Drug Abuse in Society	Geraldine Woods
Endangered Species	Clifford J. Sherry
Environmental Justice	David E. Newton
Euthanasia	Martha Gorman and Carolyn Roberts
Feminism	Judith Harlan
Gangs	Karen L. Kinnear
Legalized Gambling, Second Edition	William N. Thompson
Militias in America	Neil A. Hamilton
Native American Issues	William N. Thompson
The New Information Revolution	Martin K. Gay
Recycling in America, Second Edition	Debbi Strong
Religion in the Schools	James John Jurinski
School Violence	Deborah L. Kopka
Sexual Harassment, Second Edition	Lynne Eisaguirre
United States Immigration	E. Willard Miller and Ruby M. Miller
Victims' Rights	Leigh Glenn
Violence and the Media	David E. Newton
Violent Children	Karen L. Kinnear
Welfare Reform	Mary Ellen Hombs
Wilderness Preservation	Kenneth A. Rosenberg
Women in the Third World	Karen L. Kinnear

Books in the Contemporary World Issues series address vital issues in today's society such as terrorism, sexual harassment, homelessness, AIDS, gambling, animal rights, and air pollution. Written by professional writers, scholars, and nonacademic experts, these books are authoritative, clearly written, up-to-date, and objective. They provide a good starting point for research by high school and college students, scholars, and general readers, as well as by legislators, businesspeople, activists, and others.

Each book, carefully organized and easy to use, contains an overview of the subject; a detailed chronology; biographical sketches; facts and data and/or documents and other primary-source material; a directory of organizations and agencies; annotated lists of print and nonprint resources; a glossary; and an index.

Readers of books in the Contemporary World Issues series will find the information they need in order to better understand the social, political, environmental, and economic issues facing the world today.

BIODIVERSITY

A Reference Handbook

Anne Becher

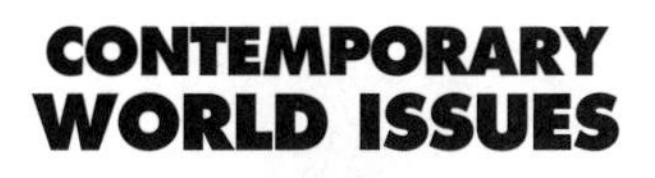

ABC-CLIO
Santa Barbara, California
Denver, Colorado
Oxford, England

Library of Congress Cataloging-in-Publication Data

Becher, Anne.
Biodiversity : a reference handbook / Anne Becher
p. cm.—(Contemporary world issues series)
Includes bibliographical references and index.
1. Biological diversity. I. Title. II. Series.
QH541.15.B56B435 1998 97-42890
333.95—dc21 CIP

ISBN 0-87436-923-1 (alk. paper)

02 01 00 99 98 10 9 8 7 6 5 4 3 2

ABC-CLIO, Inc.
130 Cremona Drive, P.O. Box 1911
Santa Barbara, California 93116-1911

This book is printed on acid-free paper ♾.
Manufactured in the United States of America

Contents

Figures and Tables

Figures

Tables

Preface

The word *biodiversity* is used with more and more frequency, but the meanings assigned to it can differ greatly. The word can be used to refer to the wealth of species existing in ecosystems throughout the world, some of great current or potential value to humanity. For others, it describes the intricate complexity of relationships tying together the species that inhabit an ecological community. This definition often includes the services that the interrelations provide to the rest of the world: the production of oxygen, the storage and cleansing of water, a laboratory for evolutionary processes, the generation of fertile soil, etc. Traditional indigenous peoples, who depend directly on biodiversity for their food, construction materials, clothing, medicines, and other necessities of life, value biodiversity not only for what it provides to them, but also as a gift from their creator. Many traditional cultures have incorporated rules and taboos that ensure the ongoing conservation of the ecosystems they inhabit.

Biodiversity is coming to the forefront of public discussion now because scientists are discovering how severely it is threatened. Recent reports show that some 40 percent of the earth's land surface has been altered by humanity. An incalculable number of species have disappeared already, and a large percentage are threatened with imminent extinction. The services we depend upon but do not always recognize will gradually cease as biodiversity wanes. Many people all over the world are concerned about this trend and are working hard to counteract it. Conservation projects ranging dramatically in size and scope are being undertaken worldwide.

The multiple issues associated with biodiversity are explained in detail in the overview chapter of this book. Readers should start with that chapter to become familiar with the many facets of biodiversity. The citations there should lead researchers to good source material, which hopefully will be available in a local library or bookstore.

The second chapter provides an anecdotal chronology of events related to biodiversity. It is designed to provide a flavor of the current events and political/scientific/intellectual climate throughout the timeline. Most of the events cited have occurred in the past two decades.

The biographical sketches in chapter three profile 23 leading names in biodiversity. Scientists, grassroots activists, journalists, a philosopher, and a sociologist are among the people featured in this chapter. Those included cannot be considered to be the main actors in the field but represent the national and professional diversity of individuals involved.

Chapter four collects a number of tables, drawings, and documents that illustrate several aspects of biodiversity. The organizational sketches in chapter five describe a host of groups that work in many ways to preserve biodiversity. Included in the contact information for each organization is its worldwide web site address, since the Internet is becoming an increasingly important place to obtain information.

Chapter six is devoted to print resources, primarily books. This chapter is divided into 13 subsections, each focusing on a different element of biodiversity. More and more information is becoming available in a nonprint format. Excellent videos and CD-ROMs are listed in chapter seven, as are databases and computer networks that provide better access to the recent surge of biodiversity-related information.

The glossary at the end of the book defines the technical terms and abbreviations used throughout the text, and the index should help locate information on any specific topic.

Researchers should remember that biodiversity is a monumental, multifaceted topic, and the tighter the focus, the more fruitful the research.

Acknowledgments

The information in this book is based on the achievements of hundreds of dedicated individuals who have been hard at work in the field of biodiversity for years. I salute them for their efforts and wish to express my hope that they will carry on their important research and activism for many more years to come.

Several people helped me with this project. Cal Roberts's networking instincts made it happen in the first place. Entomologist Eric Olson explained many current issues in the politics of biodiversity, and Joyce Gellhorn helped bounce around ideas early on during this project. Annie Lovejoy of INBio in Costa Rica was tenacious and efficient as she tracked down information I needed. Meredith Broberg provided gorgeous illustrations on very short notice. Meg Knox, Joe Richey, Ted Becher, and Ingrid Becher all read and critiqued sections of the manuscript.

My thanks to the able editors at ABC-CLIO, including Henry Rasof, Susan McRory, Kristi Ward, Allan Sutton, and Carol Estes. And finally, loving thanks to Joe, Jacob, and Flora Richey for letting me disappear from the homefront for long periods of time.

Biodiversity: An Overview

Biodiversity is a new word, a combination of the words *biological* and *diversity*. The term refers to the diversity of living beings in the world, or the great variation among the planet's live organisms. Biodiversity also refers to the interactions between these organisms that together perform many important functions that keep the planet livable. The realization that a huge number of very diverse living things exist on earth, and that the communities they form together are diversified, efficient, and productive, is nothing new. But the recent focus on this diversity by the scientific community and the general public has occurred mostly because the earth's biodiversity—its vast number of species and the interconnected communities they form—can no longer be taken for granted.

We are currently undergoing a mass extinction on the order of the one 65 million years ago that extinguished the dinosaurs and most other animals of the Cretaceous period. That wave of extinctions and the four other major extinctions were caused mostly by habitat alteration that occurred naturally, such as when tectonic plates shifted and broke apart continents, or when gigantic asteroids collided with the earth, or when dramatic shifts in temperature covered vast areas with glaciers or made them recede. The current wave of extinctions, however, is clearly the result of human-caused alteration of habitat. The human species is rapidly replacing natural habitat, whose intricate ecosystems developed over millions of years, with our own habitat (cities, farms, mines, and industry). This process leaves other species homeless. Uncountable species have died out so far this century, and estimates abound as to how many more species we can expect to lose—a species a day, a species an hour, a quarter, a third, half of all currently existing species (Lugo 1988, 59).

This precipitous loss of species is a great tragedy. Almost any ecosystem on land or sea harbors medicines, foods, resins, fibers, and organic compounds of potential use to humanity. Species living together in ecological communities provide environmental services that are vital to our survival on this planet. Mountain forests protect sources of clean water for drinking, irrigation, or generation of electricity. Wetlands are nurseries that protect young mollusks, fish, birds, and amphibians through their vulnerable juvenile stage and serve as natural filters and sinks to cleanse water and absorb it during inundation. Terrestrial ecosystems, through the decomposition of organic matter, generate fertile topsoil. Plants, especially those in dense forests and microscopic plankton in oceans, absorb carbon dioxide and produce oxygen, which balances the mixture of gases in the atmosphere. There are numerous other important functions that intact biodiversity performs to make life possible. Some of these functions have already been identified by scientists; there are probably others that we cannot even fathom.

Because of its importance to the health of the environment as a whole and to humans as a species, biodiversity is studied by a variety of fields:

- Scientists describe living organisms, their interactions, and their functions. Many branches of the life sciences broach biodiversity, including biology, botany, zoology, taxonomy, and all of the associated specialties. Some scientists take a historical approach, comparing past mass extinctions to the one we are currently experiencing. Specialists in the art and science of ecosystem restoration observe the needs and interactions of the species as they regenerate. Other specialists work on what is called *ex-situ* conservation: breeding endangered species in zoos or wildlife centers to maintain a healthy, genetically viable population in captivity until the species' native habitat is secure enough for reintroduction.

- Economists devise new economic models that factor in environmental costs of certain practices or products (petroleum, for example, should cost much more than it currently does because of the environmentally destructive methods used to extract it and its deleterious effects on our atmosphere). They describe the economic forces in the world that have led to the widespread destruction of ecosystems and propose measures that could lead toward conservation. One approach to biodiversity conservation is to assess the potential monetary value of new medicines and other useful compounds that will be discovered in intact ecosystems like forests and coral reefs.
- Political scientists and lawyers analyze environmental policies and their effects on conservation.
- Some anthropologists study indigenous peoples practicing traditional lifestyles as well as the threats they now face. Others examine societal attitudes toward nature and conservation.
- Philosophers question our tendency to assume that we rule over other species. Many indigenous societies considered themselves to be one of many species within an ecological community, all deserving respect.
- Artists, inspired by biodiversity's complexity, strive to describe its perfection in pictures, shapes, or words.
- Theologians persuade congregations to live as responsible stewards of nature.
- Pharmacists send ethnobotanists to scour jungles for substances useful to medicine.
- Agricultural specialists isolate genes in the wild cousins of our food sources in order to improve domesticated crops.

Definition of Biodiversity

Although the study of biodiversity is multidisciplinary and specialists in many fields grapple with it from their own perspectives, science claims the core of the discipline. Yet the core itself is very complex. To understand it, scientists have divided the study of biodiversity into different levels.

Three levels of diversity are generally described. The most basic is genetic diversity, or the genetic differences between individuals of the same or different species. Next is taxonomic diversity, or the diversity of organisms as divided into taxonomic categories. This level is often considered the species level, yet there are other taxonomic divisions of organisms as well: genera, families, orders, classes, phyla, and kingdoms. Taxonomic diversity is measured in sheer numbers (that is, numbers of species, genera, families, etc.) as well as how each division differs from its counterparts. Third is ecological diversity, described in terms of the organisms that live in particular communities, their interactions, and the functions these communities perform. This level describes habitats ranging in size from micro-ecosystems (such as a rotting log) to landscapes (such as watersheds—geographical regions that extend from water sources to river drainages). The diversity at each of these levels is equally important, for the health of each level is bound to that of the others.

Genetic Diversity

This level of diversity refers to the variety of genes available to a taxonomic division such as a species. Genetic richness is important for individuals as well as for a species as a whole since genetic makeup determines, among other things, body size and form and resistance to disease, poisons, or such hardships as drought. If a particular species or a population of a species (a group of individuals living in a geographically separate area from others of the same species) is numerous, its members descended from unrelated parents, the species is likely to be genetically varied or rich. Individuals with traits that give them the best chance for survival will reproduce more successfully, and over many generations their genes will come to dominate. If a species remains genetically rich, it will continue to evolve in response to changing environmental conditions.

A species is genetically poor if its numbers are small and most members are related to one another. There is not as much genetic variation, which is to say that individuals do not exhibit unique traits or survival strategies. This condition, together with the lack of opportunity to flourish in its territory (if, for example, its habitat is being destroyed), will weaken the species. Genetic poverty is generally a precursor to extinction.

Taxonomic (Species) Diversity

This second level of diversity concerns the differences between taxonomic divisions such as species. Species are defined roughly as groups of like organisms that mate among themselves but not with others. About 1.4 million species have been identified so far. When pressed to give an estimate of how many species may inhabit the planet earth, biologists respond with numbers of between 5 and 50 million. The vast disparity among estimates is due to differing hypotheses of how many undiscovered species might exist. Most of the larger species have been identified already: mammals, reptiles, amphibians, birds, trees, etc. The smaller organisms, however, especially insects, fungi, bacteria, and viruses, could be far more numerous than anyone suspects.

Taxonomic diversity can be described in ways other than counting species: the size of their populations within their habitat (whether they are common or rare); the size of the area they inhabit (the whole world or a narrow mountain valley on a single island); the variety of places they are found (adaptability to climatic and geographic conditions); how much they vary from other species (do they have close relatives?); how hardy or fragile they are; and what roles they play in their habitat (as pollinators, predators, providers of shelter or food, etc.).

Body size, range, population size, uniqueness: none of these characteristics alone makes any species more important than another. Each species is invaluable for the information in its genes and for the knowledge gained from watching it at work in its habitat. Most important may be how an organism interacts with others. Each species has developed intricate interdependencies with others in its ecological community. Certain *keystone* species—named for the stone Romans placed at the top of their arches to prevent them from collapsing—can be considered the most important species in their ecosystems. If the keystone species in a community becomes extinct, it will lead to many more extinctions in that community. But the keystone species in most communities has not been identified. The overall health of a community may depend on an infinitesimally small organism not even visible to our naked eye.

Ecological Diversity

The third division of biodiversity refers to ecological communities. These come in a range of sizes. The ecosystem level is the one most commonly mentioned by biologists when describing ecological diversity and is defined as a habitat with certain physical characteristics like elevation, temperature, moisture, wind, weather, soils, and all of the life-forms within it. A step up from ecosystems are landscapes, which are the conglomeration of many ecosystems in a defined region such as a watershed. Groups of landscapes make up biomes, and the largest ecological community is the biosphere itself: all of life in the thin envelope between the earth's crust and the atmosphere.

When scientists study ecological biodiversity, they consider far more than the species themselves. Important to the concept of ecological community are the interactions among species, and the relationships between species and the habitat's physical characteristics. As species interact among themselves and with their physical surroundings, they set in motion certain cycles that keep vital elements moving throughout the biosphere. Hydraulic cycles (how water moves from the atmosphere to the land to the ocean and back into the atmosphere) depend upon intact forests, wetlands, bodies of water, and all ecosystems in between. Plants and animals are vital for cycling the building blocks of life: nutrients like carbon, nitrogen, sulfur, and oxygen.

The image of communities side by side and nesting within larger ones illustrates their interdependence. Contamination in one ecosystem can damage another: chemical runoff into a river can poison a wetland downstream, and deforestation of a mountain watershed can fill a river with excess sediment and cause floods downstream. Landscapes that envelop seriously altered ecosystems suffer a loss of biodiversity as a whole, because many species migrate between neighboring ecosystems. Landscape-level perspectives are important, especially for conservationists and wildlands managers. Their mandate to conserve species cannot be accomplished by protecting one small ecosystem. What happens nearby is also of crucial importance.

How Biodiversity Evolved

As useful as it is to divide biodiversity into its three levels, it is also helpful to understand how the 5 to 50 million species currently living on earth came to be. Why such diversity? How did it all begin?

Although we don't know for certain how life originated, scientists posit that when certain fortuitous combinations of chemicals were exposed to phenomena like lightning or volcanic activity, sea-based microbes appeared. Once they formed 4 billion years ago, they immediately began evolving in response to environmental conditions. By 2 billion years ago, the microscopic, one-celled creatures floating in the seas had developed the capacity to harness the sun's energy for photosynthesis, which gave them more energy for growing and developing. It wasn't until 570 million years ago that more complex organisms began to graze on simple algae, thereby establishing the first link in the food chain.

In addition to helping the photosynthesizing organism itself, photosynthesis benefited future life-forms. One of the by-products of photosynthesis is oxygen. By 450 million years ago, photosynthesis practiced on a massive scale had released enough oxygen into the atmosphere to form a layer of ozone that shielded the earth from the sun's powerful rays. With this protective filter in place, the land surface of the planet became a more hospitable habitat, and marine organisms began emerging from the oceans (Erickson 1991, 28–29). Simple algae and lichens colonized dry land and eventually formed the base for soil, which allowed flowering plants to take root. Eventually, marine animals crawled out to graze on terrestrial plant life. Slowly, over the eons, organisms in the sea and on land evolved into the plethora of forms that characterize biodiversity today.

The Rise of Biodiversity

To survive and thrive, organisms need to solve a few basic problems: what to eat, where to live, how to find protection from enemies, and how to reproduce. Biodiversity arises from the unique way each organism responds to these fundamental questions of existence. Organisms don't respond to these needs in isolation; other organisms help.

Mutually beneficial interactions between life-forms are the basis for biodiversity. Algae and lichen form the soils that allow plants to root on land, and plants drop leaves that feed algae and lichen. Fungus and bacteria in the soil anchor on plant roots, channeling nutrients from the soil into the roots. In return, they absorb excess sugar generated by the plant through photosynthesis. Most flowering plants must be fertilized during their flowering in order to produce seeds. Many plants lure pollinators to help them by sporting colorful or aromatic flowers and exuding enticing nectars. Scent-driven pollinators like bats or flies are attracted to sweet-smelling flowers that may be dull in color. Flowers that are pollinated by keen-eyed insects are brightly colored. Syruplike nectar draws butterflies and bees. Each pollinator is perfectly suited to help a particular plant; each plant offers its pollinator exactly the right type of food. The next stage of plant reproduction involves dispersal of the plant's seeds, and in a similarly effective strategy, plants offer to the right dispersal agent—bird, squirrel, monkey, cow—a neat packet (such as a berry or fruit) containing a seed that the disperser can regurgitate or defecate along its path. Some seeds will not germinate unless they have passed through the digestive system of the appropriate animal!

Nature can be a rough and dangerous zone, and species also must protect themselves from enemies in order to survive. A great variety of protective strategies differentiates and diversifies organisms. Some animals are swift, able to flee by land, air, or water from their predators. Those animals have highly developed senses and are able to see, hear, or smell their enemies from far away. Plants, insects, and amphibians such as poison dart frogs manufacture chemical compounds that make them distasteful or even poisonous to their enemies. Other species that do not manufacture these chemicals mimic those that do. Some tasty plants such as cacti are covered with spines to deter tender-mouthed predators. Protective patterns or colorings allow plants or animals to blend into the background. Some plants do not cluster together but instead occur at a great distance from one another so that predators are less likely to locate their progeny. These are only a few of the myriad defensive strategies that plants and animals have developed over millennia.

Mirroring each defensive strategy, predators have developed equally sophisticated offensives. Extreme patience allows frogs to remain absolutely still until a fly comes close, or jaguars to lie in wait for deer. Certain predatory insects and birds have developed a resistance to poisons, or a way to deal with spines or other discomforts. Some predators mimic other species that are harmless to their prey.

Evolution

The match between predator and prey, between plant and pollinator, is driven by a process called evolution. For billions of years, organisms have developed through evolution, which means they have adapted to prevailing conditions and the resources available. There are two major forms of evolution. In the first, a species gradually changes until it becomes quite different from its original form; this is called vertical evolution. In the second, one species branches into two or more species, each best suited for certain environmental conditions. This is called adaptive radiation.

A well-known example of the vertical type of evolution is that of the *Biston betularia* or peppered moth. Collections made of this moth in very polluted areas of Detroit and England in the 1950s show that about 90 percent of the individuals were very dark—an advantage for survival as their black wings matched the black industrial sky. After England and the United States passed stringent pollution control laws, however, the skies became clearer. Collections made in the same locations in 1995 showed that only 20 percent of *Biston betularia* individuals had dark coloring (Yoon 1996b). The species had evolved in response to the environment by developing protective coloring. The gene pool of the U.S. and British populations of *Biston betularia* was rich enough so that as air quality improved, individuals with genes favorable for the newly clear skies thrived. Lighter-colored individuals were more successful in reaching maturity and mating. The genes they passed to their offspring gave them the same favorable trait: light-colored wings. Over time, individuals with genes for the newly favorable trait came to dominate in the species. This process is called natural selection. Natural selection can continue until a new species develops and the original, more primitive one dies out. The brainy *Homo sapiens* became so different from its *Homo erectus* ancestors that it ceased to mate with its primitive relations and might even have engaged in warfare that killed off *Homo erectus*.

In the evolutionary process known as adaptive radiation, a group of individuals of the same species is isolated from other members of the species. The isolated population responds to its environment through natural selection and slowly grows so different from other populations of its species that even if the different populations were to meet, they would no longer be able to mate and reproduce. Through adaptive radiation, one species can branch to become two or more.

A famous example of this mode of evolution created the species known as Darwin's finches. When Charles Darwin, the father of evolutionary theory, arrived on the Galápagos Islands west of Ecuador, he found an array of endemic finch species descended from a small population that had landed on the islands 4 million years ago. Their offspring had branched into 14 species, each filling a separate niche (a space in the physical or organizational structure of an ecosystem). Each finch species is different from the others in minor but crucial ways. Some feed primarily on seeds and have triangular, hard beaks designed for crushing. Others eat insects; one has a long, thin bill to probe tree bark and pry out bugs. Another even pecks at a larger bird and sucks its blood. In another place with many orders of birds, these niches would have been filled by very specialized families and species.

But with so few birds living on the islands to begin with, one species fanned out to do it all. It was this example of adaptive radiation that convinced Darwin that evolution was the vehicle for the proliferation of species.

Speciation usually requires tens of thousands to millions of years. Scientists were recently surprised, however, by findings that the 300 species of cichlid fishes in Lake Victoria in East Africa evolved in only 12,000 years. Apparently cichlids are weak swimmers and do not stray far from customary habitats. If they lie in a sandy-bottomed area, for example, they may never cross a narrow, rocky corridor to reach the next sandy area. Small populations of cichlids in Lake Victoria quickly grew so different from the rest that they formed distinct species. The discovery of the cichlids in the 1990s revealed one of the most dramatic examples ever of rapid speciation (Yoon 1996a).

Another process leading to today's biodiversity is immigration. Each form of life probably originated in one place on earth, but then for a variety of reasons, some individuals moved. Some insects and seeds are designed to catch wind and fly; they become what biologists call aeolian (wind-borne) plankton (Wilson 1992, 20) and colonize any place they land. Birds can be caught up in storms and carried far from their homes. Other vertebrates have been known to swim or raft the ocean and make a new home where they beach. A period known as the Great American Faunal Interchange occurred about 3 million years ago when the Panamanian land bridge emerged from the ocean. Northern placental mammals moved to South America and southern marsupials headed north. Some immigrants and some natives dominated on each continent, crowding out others. Adaptive radiation led to the creation of many new species. The result is a mixture of placentals and marsupials throughout the Americas: 50 percent of South America's mammals are placentals, North American in origin, and 20 percent of North America's mammals are marsupials, originally South American (Raup 1991, 133–134).

Evolution has not proceeded in a smooth upward line from microbe to man. It has been punctuated by huge extinctions that have wiped out between 70 and 95 percent of the species inhabiting the earth. There have also been periods of extremely rapid development of new species, usually after mass extinctions, once conditions favorable to most forms of life returned. These periods are probably due to vacant niches left by the extinct species. For example, after the dinosaurs disappeared, there was an abundance of food (the vegetarians consumed huge amounts of plant material) and fewer predators (the tyrannosaurus and other carnivores had disappeared as well). The vacant niches in the ecosystems of the day allowed for the explosive evolution of mammals.

Measuring Biodiversity

Billions of years of evolution and extinction have yielded between 5 and 50 million species currently inhabiting earth—only 2 to 4 percent of the number of species that have existed since the beginning of life. Counting species, or any of the other taxonomic divisions that scientists use to classify life-forms (genera, families, orders, classes, phyla, and kingdoms), is one way to quantify the planet's biodiversity.

Scientists have identified and named only about 1.5 million species, but they hypothesize that there are millions more. The large, visible species of mammals, birds, and trees are probably mostly accounted for, although on rare occasions a new one is discovered. Recently, a brand new species of deer was found in an isolated Asian valley. The species that scientists believe to be most numerous are the ones that are hardest to spot. Some species are invisible to the naked eye (viruses, bacteria, protozoa—one gram of soil can contain 10 billion bacteria). Others (such as beetles and other insects) are larger but live in cracks at the tops of trees and other places that people have a hard time reaching.

In 1982, Smithsonian Institution entomologist Terry Erwin performed a survey of insects inhabiting the canopy of a Panamanian rain forest and found that a single tree could harbor more than 150 species of beetles—and that those 150 beetles lived on no other species of tree. He made some calculations and estimated that if he could generalize his observation that species such as beetles were host-specific (that is, they lived only on a certain species of tree), there might be as many as 30 million species of arthropods (insects, spiders, etc.) living in tropical rain forests (Wilson 1992, 137–140).

Species-Rich Ecosystems

Tropical rain forests, which comprise only 7 percent of the earth's surface, harbor between 50 and 90 percent of the earth's species. A recent study in the Atlantic coastal forest of Brazil found 476 species of trees in a one-hectare plot (two and a half acres). This compares to an average of 2 to 20 tree species in a North American plot of the same size (Associated Press 1996). Entomologist E. O. Wilson collected 43 species of ants from a single tree in the Tambopata Reserve rain forest in Peru; in all of Great Britain there are only about 43 species of ants (Wilson 1988, 9).

Why are tropical rain forests so much richer in number of species than other habitats?

Certain conditions are especially conducive to most life-forms. Abundant moisture, warm temperatures, and sunlight gave rise to the original sea-based microbes 4 billion years ago, and these conditions prevail in the tropics. Their location on the globe endows them with year-round sun and warm temperatures. The number of species grows steadily as one traces a line from the poles (where the sun hides for half of the year and year-round ice caps keep temperatures cool in summers) to the equator (where the sun beams down all year). Heavy rainfall year-round in a tropical rain forest allows more species of trees to grow, and trees grow higher because they can continue growing during the entire year. Further north or south, in temperate regions that have cold winters, trees virtually hibernate during the winter. Their yearly growth cycle is half as long as that of their tropical relatives.

The diversity and height of trees in the tropics is important because trees are the frames of the forest, supporting many other organisms. Rain forest trees are draped with other plants. Vines sprout leaves high enough to catch the sunshine near the top of the canopy, then shoot cablelike roots sometimes hundreds of feet downward, curling around tree trunks or plunging straight down to the forest floor. Epiphytes grasp trunks and branches with threadlike roots and gather moisture and nutrients from rain as it drips down through canopy layers.

This multilayered rain forest framework hosts uncountable mini-ecosystems. Water-storing bromeliads (air plants that anchor onto tree limbs, for example) are distinct mini-ecosystems in themselves and enjoy a mutually beneficial relationship with tree frogs. The female vermilion poison dart frog lays eggs in damp humus on the ground, but soon after the eggs hatch, she carries them to her "nursery," a tiny pond formed by the cupped leaves of a bromeliad high in a tree. Here they develop unthreatened for two months. The female frog deposits protein-rich unfertilized eggs in the water, which the tadpoles perforate and suck for nourishment. Their waste products decompose in the water and provide the bromeliad with absorbable nutrients (Blake and Becher 1997, 25). Many insects in their larval stage also inhabit these tiny ponds, nourishing the bromeliads as they take shelter in them.

The many niches within a tropical rain forest allow for a degree of speciation that would be impossible in a simpler ecosystem, where the physical and organizational structure is simpler and species must remain flexible to survive with a food supply that varies throughout the year. Highly biodiverse ecological communities have tightly packed guilds (groups of species with similar behaviors). The existence of so many potential competitors for food, pollinators, or shelter encourages specialization (Terborgh 1992, 65–70).

The ten species of antwrens living in Amazonian forests exemplify a tightly packed guild. These small, insect-hunting birds are similar in form and appearance but inhabit different levels of the forest and use specialized food-hunting strategies. The complexity of the rain forest ecosystem offers them opportunities to scavenge in dense foliage, leaf litter, or along bare branches, at four different levels of the forest. Four of the antwren species that hunt for ants exclusively on live leaves inhabit different altitudes in the forest. Such extreme specialization has allowed them to avoid competing with one another (Terborgh 1992, 70–71).

Plants, too, can avoid competition through specialization. A year-round growing season in tropical forests allows different species of plants to avoid vying for the services of a common pollinator. Each of the many plants using the same pollinator in a Costa Rican rain forest flowers at a different time of year. The plants are all able to reproduce, and the pollinator has a yearlong nectar supply (Pimm 1991, 205).

Other Measures of Biodiversity

For their astonishingly high number of species, tropical rain forests are usually labeled the most biodiverse of all systems. However, the number of species in a given place is only one of several ways to measure biodiversity. It is also useful to count how many upper taxa (kingdom, phyla, class) are found in a particular ecosystem. This number represents a more profound diversity, since organisms in these groups are not as similar to one another. Another useful measure is the number of endemic species, species found nowhere else in the world. The presence of large numbers of endemic species increases the importance of an ecosystem from a scientific perspective. If an endemic species disappears from its ecosystem, it can never be studied or appreciated anywhere else. A further measurement of the value of an ecosystem can be made by tracking its visitors. Many ecosystems host animals for only a short time each year, but migrants may depend upon those ecosystems to successfully reproduce and raise their young. Ecosystems that host migrants are of particular importance to the overall health at the landscape or biome levels.

Diversity of Upper Taxa Marine biologists are quick to decry a species-based measure of biological richness, because oceans are less species-rich than many terrestrial ecosystems. But oceans, they point out, are home to members of 43 of the 70 phyla recorded, double the number of phyla with representatives on dry land (Ray 1988, 39).

Endemism Endemic species usually occur in isolated ecosystems, most commonly oceanic islands. Endemic species on islands are descendants of stragglers that either immigrated to an island or lived in an area that broke off and became an island. Through adaptive radiation, the descendants evolved into distinct species. The Galápagos Islands host a high number of endemic species. Madagascar has the most diversified group of endemic primates: 29 species of lemur have evolved in isolation since the island separated from Africa 200 million years ago (Mittermeier 1988, 150).

Other ecosystems rich in endemism are isolated not because they are surrounded by ocean and are far from similar ecosystems but because they are separated by barriers that cannot be crossed by most species. Lakes are one such type of ecosystem, and Lake Victoria's cichlids are a prime example of endemism. Mountainous terrain defined by high ridges and deep valleys forms isolated ecosystems with relatively immobile endemic species. Each of the six species of *Partula* land snails of Tahiti inhabits its own volcanic valley. As the crow flies, the valleys are minutes away. But as the snail crawls they are impossibly far from one another. This is how the *Partula* speciated: individuals who somehow arrived in one valley or another never regained contact with the rest of the population and evolved, through adaptive radiation, until they became so different that they were classified by taxonomists as separate species (Gould 1993).

Roles of Ecosystems as Measures of Richness Another measure of an ecosystem's biodiversity concerns which species visit the ecosystem and for what purpose. Inland marshes are important because many species of waterfowl rest and replenish their energy supplies there during yearly migrations. The mangrove forests of coastal wetlands have unusual root systems. At low tide, mangrove roots remain above water and at high tide they are covered by a brackish mixture of fresh and salt water. A web of buttress roots anchors the tree trunk so that the force of the incoming or outgoing tide won't dislodge it. The dense root mass provides a safe habitat for spawning female fish and their offspring, since most predators cannot penetrate the tangle. As the young grow and venture into the more open area of the wetland, they are still protected, because the salt content of the wetland water varies more than the ocean and most predators from the open ocean cannot adapt to it. Although many species spend only a small percentage of their lifetimes in a coastal wetland, this habitat is crucial to their propagation and hence to the maintenance of biodiversity.

The Importance of Biodiversity

Valuing biodiversity for its own sake is a natural approach for biologists and conservationists. Most other people, however, consider biodiversity in terms of what it offers humanity. Decision makers, from politicians to individuals making everyday decisions, think of themselves and their fellow humans first. Why should people care about the extinction of a species that may be important for its ecosystem but has no obvious usefulness for us? (Recall the controversy about the spotted owl and the snail darter.) Why care about ecosystems that seem alien or unpleasant—swamps, for example?

Importance of Biodiversity in Ecological Communities

Cities and towns, where most of us live, are usually built on the site of greatly altered ecosystems. We pave roads, cement sidewalks, and surround ourselves with the kinds of plants we favor: grass, shade or fruit trees, and flowers. Although remnant patches can still be found in vacant lots and train yards, the original vegetation is largely gone. But despite our ability to alter the original communities in our immediate vicinity, we still live within a larger ecological community. The health of neighboring ecosystems is vitally important for our survival.

Wetlands, for example, are among the most threatened ecosystems within human-populated landscapes. They are not particularly hospitable to humans. They can be smelly and are often full of biting insects and even dangerous animals like water snakes. Since cities are always in the process of expanding outward, they often overtake wetlands. Developers commonly drain them, or fill them in, and build on top of the "reclaimed" land.

The developer's profit and the city's new space may seem like great gains. But the conversion of a wetland impacts more than the fish and mollusks that use the wetland as a nursery and the birds that eat them. It also implies a loss of the important services the wetland provides to its neighbors. Coastal wetlands buffer inland areas from tidal surges during ocean storms by absorbing excess water. Wetlands absorb excess river flow and rainwater during floods. Their unique mixture of flora and fauna, together with the cycling of water within the habitat, purifies water contaminated by sewage and industrial pollutants. Water purified by wetlands may flow downstream or into an aquifer, to be consumed by other humans.

Altering landscapes upriver from inhabited areas can also induce undesirable consequences. If vegetation alongside the headwaters of a river is removed, the river can become an uncontrollable torrent during storms, causing catastrophic floods. Intact vegetation absorbs heavy rain much more effectively than denuded land. A forest serves as a filter, each layer of the canopy absorbing some of the water before the rain reaches the spongy earth below. When heavy rains drench treeless farms, wasteland, or urban streets, the land is hit suddenly with more water than it can absorb, and rivers may swell beyond the capacity of their beds.

Hurricane César, which devastated Costa Rica in July 1996, was a sobering rejoinder to too many decades of rain forest destruction. The rain fell heaviest on the steep southern Pacific slope, where most of the original rain forest had been cut decades before for cattle grazing and agriculture. Mud slides destroyed homes and killed their inhabitants, and entire towns were washed away by the disastrous floods. The storm would not have wreaked such damage had the hilly forest ecosystems been intact.

In addition to the roles that ecosystems play in maintaining the health of larger ecological communities, ecosystems are the heart of the global system that recycles the chemical elements that make up our soils and atmosphere. Ecosystems cycle nutrients needed by all forms of life, such as oxygen, methane, carbon, sulfur, and nitrogen. Nutrient cycles involve the action of plants, their consumers (animals), and decomposers (such as fungi and bacteria). Plants convert sunlight into sugar, via photosynthesis, which provides them with the energy to grow. The biomass they produce—leaves, flowers, fruits, etc.—is eaten by such consumers as insects and larger animals, which excrete nutrient-rich feces that fall to the ground and are broken into chemical elements by fungi and bacteria. These elements can then be absorbed by plants and used to fuel the photosynthesis by which more plant matter is produced. When plants die, they too are broken down into their original chemical elements and absorbed by other plants.

Nitrogen is one nutrient cycled in this process. Nitrogen can be captured from the air by the roots of certain plants and is converted into protein in their foliage or fruit. Animals eat the protein. The protein returns to the ecosystem as the animal defecates or when it dies and decomposes. (It returns in a simpler state when the plant dies.) Bacteria and fungi break down proteins and, in a complex process, convert them back into nitrates, which can be used again by plants. Unfortunately, this cycle is easily interrupted. When humans take plant or animal proteins from the place they originated and consume them elsewhere the ecosystem's nitrogen is not naturally replenished. In a cultivated field, this necessitates the application of fertilizer containing nitrogen. If a forest is cut down and some tree parts are burned or left to rot, the nitrogen level will be high for a short time and the land will be fertile for agriculture or grazing. But since nitrogen cannot continue to cycle, the land will soon become infertile, and what was once a productive ecosystem is rendered barren.

Living organisms have adapted to the conditions presented by a very precise mixture of oxygen and carbon dioxide in the atmosphere. Animals breathe in oxygen and breathe out carbon dioxide. Plants absorb carbon dioxide and produce oxygen. Foliage-rich forests and oceans are called carbon dioxide sinks because they are such effective absorbers of carbon dioxide. Phytoplankton—tiny photosynthesizing organisms that float in the sea—are responsible for the great bulk of the oxygen–carbon dioxide exchange. These organisms—so minute that scientists didn't know they existed until very recent innovations with electron microscopes revealed their presence in seawater—cycle huge quantities of the gases. Warm temperatures cause the plankton to work faster; the oxygen they release cools the air. The response of this cycle to climatic conditions led Drs. James Lovelock and Lynn Margulis to formulate the Gaia Hypothesis of the earth as a self-regulating organism. According to the Gaia Hypothesis, when the temperature of the atmosphere rises to temperatures above the optimum for life on earth, plants and plankton will photosynthesize faster, releasing more oxygen than usual, which will cool the atmosphere back to optimum temperature. If the atmosphere is too cool, reduced photosynthesis will swell the quantity of warming carbon dioxide.

As cyclers of nutrients and chemicals, ecosystems are inherently productive. Working together, all of the inhabitants generate food for themselves and usually surplus as well. A temperate grassland, for example, is a system made up of many elements: a collection of grasses; insects and birds that feed off the grasses, pollinating them and dispersing their seeds; mammals that eat plants and animals and fertilize the soil as they defecate; fungi and bacteria that inhabit the soil and break down feces and dead plants and animals into elements that the grasses can absorb through their roots. The grassland ecosystem produces food for each of its inhabitants, and the surplus can be culled for non-inhabitants—i.e., people—who harvest wild grains or hunt wild animals.

Importance of Biodiversity at the Taxonomic Level

If any of the species inhabiting the grassland described above were eliminated from the picture, the ecosystem's productive capacity would plummet. If a species of grass were to become extinct, its consumers might, too. If a pollinating insect died off, the plant it pollinated might not reproduce. Without a mammalian grazer, the grass might overgrow other plant species and choke them out. If the mammalian grazer lost its predator, it might reproduce in great numbers, decimate its food source, and self-destruct. If degraders like fungi and bacteria were to disappear, dead bodies and plants would pile up quickly. Scientists have documented numerous cases of ecosystems losing a key member and collapsing.

The diversity of species is what keeps ecosystems functioning. All species have a variety of functions in an ecosystem. Besides being important decomposers of waste products and dead plants and animals, bacteria and fungi live in digestive tracts and help some species of animals digest food. Plants are the base of the food chain, producing protein and sugars in the form of leaves and fruits. They depend upon some of their animal clients to help them reproduce: insects and birds often pollinate flowers, and fruit-eaters ingest seeds, depositing them elsewhere when they defecate. (The seeds of some plants will not germinate unless they have passed through the gut of a certain animal.) Animals need food, which comes in the form of a plant or another animal. The web of interdependence woven by species living in a particular ecosystem can be so intricate that scientists only begin to unravel its minute particulars after a lifetime of study.

Keystone Species

So there are usually no easy answers to the question "Are certain species more important than others?" It depends upon how "important" is defined. Each ecological community has certain keystone species whose role is so important that the community would collapse without them. Unfortunately, most keystone species have revealed themselves only after their disappearance. The sea otter, which inhabited the kelp beds off the Pacific Coast of North America, was hunted into virtual extinction during the last century. Otters eat sea urchins. Unthreatened, sea urchin populations exploded, devouring the shallow sea floor's kelp and other seaweed beds. That once lush habitat turned into a barren waste because the keystone of the ecosystem had been removed. Fortunately, conservationists were able to bring back the otters, which had survived at the extreme north and south margins of their range. As the otter population was restored, urchin numbers shrank and kelp beds regained their former luxuriance (Wilson 1992, 164–165).

Another keystone species is the Everglades alligator, a predator of the gar fish. When alligator populations diminish, gar fish multiply out of control and decimate their prey, and the shock waves shake the food chain to its base. Alligators also dig "gator holes," deep depressions in the wetlands that are fertilized with leftover food and alligator excrement. In times of drought, gator holes become refuges for aquatic species, which stay there until the wetlands fill with water again. Everglades alligators also build large dirt mounds for their nests; every year they add more soil, and eventually trees take root in these mounds. Herons and egrets nest in the tree branches, where their young are protected from predators by the nesting gator below (Durrell 1986, 120).

Invertebrate animals (e.g., insects) are keystone phyla for the global ecosystem. They perform many crucial functions. They are indispensable links in the food web, they maintain the structure and fertility of the soil, they cycle nutrients, they act as pollinators, they disperse seeds, they create physical habitats (in the case of corals), they regulate potentially harmful organisms, and they eliminate waste (Stevens 1993).

Although we know that keystone species exist, their identities, as well as the conditions they require, often remain unknown. Bacteria and fungi may be among the most delicate. We don't know what they need in terms of nutrient combinations or moisture levels or temperatures, but we do know that their role in breaking down nutrients in soil so that plants may absorb it is absolutely essential for our survival. Interfering with ecosystems, or with species within ecosystems, can be a deadly game. In their book *Extinction,* Paul and Anne Ehrlich use a metaphor to describe the folly of tampering with ecosystems. Species are equated with rivets on an airplane, and human actions leading to their extinction are technicians who are popping off the rivets one by one. At some point—and no one knows when—enough rivets will have been popped for the plane to collapse.

Other species may not play a keystone role while the ecosystem is healthy and functioning but can rise to save it in crises. One yeast that is normally very rare in aquatic ecosystems reproduces abundantly when mercury levels become high due to natural causes or industrial contamination. The yeast digests the mercury and cleans the ecosystem. When the mercury level is reduced, the yeast population declines again (T. Lovejoy 1995, 85).

Human-Centered Measures of Species Importance

Every species is of value to its own ecological community, but many also have direct, tangible uses for humanity. People have hunted and gathered their food in the wild for thousands of years, and most societies have domesticated the species they have found to be useful. An ever-evolving array of wild species assures genetic reservoirs for domesticated crops, and new discoveries continually affect our diets.

In addition to furnishing our food, biodiversity provides us with medicines. People who live close to the land use herbs to cure stomachaches, headaches, high blood sugar, heart problems, etc. The world at large has already adopted some of these. Quinine from the cinchona tree prevents malaria. Rosy periwinkle from Madagascar cures childhood leukemia. Willow bark—from which aspirin was synthesized—soothes headaches. About one-quarter of the medicines prescribed in the United States contain active ingredients derived from plants (T. Lovejoy 1995, 83).

In biodiverse ecosystems where predators abound, plants have developed a variety of chemical compounds to repel them. These compounds will often work for humans as pesticides. The British Technology Group discovered that the tree species *Lonchocarpus costaricaensis* found in Costa Rica was a powerful nematocide, or nematode (roundworm) killer. Since nematodes are major threats to potatoes, tomatoes, bananas, and other crops, this discovery was particularly useful (A. Lovejoy 1996). Also, certain fungi may help control locusts and grasshoppers

(T. Lovejoy 1995, 82). These natural methods are much safer than the toxic pesticides that farmers frequently use.

The Importance of Genetic Diversity

Compelling examples of what happens without a rich genetic pool are found readily in agriculture, where genetic diversity is intentionally bred out of major crops and the diverse species of the natural ecosystem are replaced by a single species planted over vast areas. Crop varieties are selected for certain qualities: high productivity, ease of harvesting, attractiveness of the product, resistance to spoilage. But sometimes unexpected occurrences make this monocropping backfire. Blight can strike, or an unusual predator, or a mold. Because modern agriculture consists of huge plots of just one crop, plagues or pests can decimate a region's entire harvest. The genetic variation in a species that had evolved naturally would increase the probability that individual plants would be resistant to common plagues. And in a natural ecosystem, such a vast monocropped expanse would not exist to fuel the reproduction of the plague. But without genetic diversity, the crop is very vulnerable.

To prevent scenarios such as these, a number of institutions have devoted themselves to bolstering genetic reserves for the crops the world's population depends on most: rice, wheat, corn, potatoes, sugarcane, and many more. Institute researchers seek out rare varieties developed and tended by traditional farmers, and the wild relatives of those species, and test their special qualities. A barley from Ethiopia furnished a gene that now protects California's barley crops from yellow dwarf virus; Asian rice has been fortified by genes from an Indian rice species that protects it from the major rice diseases; a wild Peruvian tomato has sweetened domesticated tomatoes in the United States (Plotkin 1988, 110). Botanists have discovered a variety of the tomato from the Galápagos Islands that is salt-resistant—a potentially important trait for plants grown in coastal areas where farmland is becoming increasingly salinized (*Sinking Earth* 1987). In one famous instance, a rare species of perennial corn discovered in Mexico in 1977 proved resistant to the seven major diseases of corn. It grew in the wild on only one 6-hectare plot, but its potential was so great that a huge 135,000-hectare biosphere reserve was established by the Mexican government and UNESCO's Man in the Biosphere program (Iltis 1988, 103). In this instance, a single species inspired the protection of an entire region. Indeed, protecting ecological communities as sanctuaries for valuable genes is one of the most compelling arguments for conservation.

As the importance of genetic diversity and connections between species that form ecosystems becomes common knowledge, one might expect people to adopt practices that would protect this invaluable biodiversity. Unfortunately, species are disappearing so quickly that we can consider ourselves to be living in a period of mass extinctions comparable to those of ancient times—each of which claimed up to 96 percent of the species then existing (Raup 1988, 52).

Threats to Biodiversity

Global biodiversity is currently in great danger. Natural habitat, with its ecological communities, species, and genes, is being destroyed rapidly. As with estimates of how many species exist, estimates of how quickly species are disappearing vary widely—from one species per day to one species per hour; between 15 and 50 percent of tropical species by the end of the twentieth century (Lugo 1988, 59). But even the most conservative estimates indicate that we are in the midst of a crisis on the order of the great prehistoric extinctions. The difference lies in the cause of the extinctions. In ancient times, geological shifts and natural catastrophes were the culprits. Today's crisis is caused by humankind.

Causes of Extinction in Ancient Times

The main factor that brought about prehistoric waves of extinction was the alteration of habitat by natural forces. Geological changes caused gradual shifts that slowly altered habitats. As continents collided, the floors of shallow seas were thrust above water level. During the most massive extinction crisis recorded, 250 million years ago, 95 percent of marine species died (Frankel and Soulé 1981, 23–24). The shallow continental sea slope habitat, with its coral reefs and sea grass colonies, has always been one of the most species-rich habitats. After each of the massive sea life extinctions in ancient history, it took 10 million years for coral reefs to regain their complex magnificence.

In addition to continental drift, geological phenomena such as catastrophic earthquakes and volcanic activity caused major extinctions. One widely accepted hypothesis for the demise of the dinosaurs is that a meteorite hit earth and sent up a huge, dense cloud that darkened the sky for a decade. Plants, the base of the food web, could not photosynthesize without sunlight, and nearly all life-forms were extinguished. Floods also caused some extinctions. Natural events like these altered landscapes very quickly—too quickly for most species to adapt to the changed landscape.

Climatic change has also altered habitat. Usually it occurs in cycles, as the earth's axis tilts closer or further from the sun, or due to variations in atmospheric composition.

A warmer climate was one factor that scientists think conspired in the extinction of the woolly mammoth and many other species in the late Pleistocene period (about 13,000 years ago). Mammoths were adapted to cool areas, and so as the climate warmed their habitual grazing grounds, they were forced to migrate northward to cooler temperatures. But closer to the pole there is less vegetation, and in winter months daylight is short. There was simply not enough vegetation or enough time for the great beasts to consume as many calories as they needed to survive (Frankel and Soulé 1981, 24).

Humans play a part in some theories of prehistoric extinction. Some scientists believe that the mammoths may have been hunted to extinction, along with other Pleistocene megamammals like the stag moose. And recently, Drs. Preston Marx and Ross MacPhee have proposed that mammal disappearances in the late Pleistocene were caused by diseases carried by the humans and their pet dogs who had recently crossed over the Bering Strait from Siberia. Marx and MacPhee are only now testing the mummified remains of Pleistocene mammals for the presence of disease pathogens (Stevens 1997). The hunting and disease theories garner more strength in a scenario that includes a shrinking habitat. It is easier to hunt animals to extinction if they are confined to a small area. And the stress caused by a disappearing habitat weakens animals, making them less able to flee hunters and more vulnerable to disease.

Events such as sudden geological catastrophes, gradual climate change, and continental drift were major factors in the elimination of species in ancient times; disease and overhunting could have been a secondary cause at some periods in history. The gradual changes to our topography and climate are still occurring, and the sudden ones can occur at any time. Disease epidemics rise and wane over the years. But the main difference between threats to biodiversity in ancient times and in the present is that today humans are the primary agents threatening biodiversity.

The Current Biodiversity Crisis

Ironically, learning how to alter habitat—burning grasslands to scare up game and ease the hunt, cutting patches of forest for cultivation, and diverting water flow for irrigation—helped humans in their early struggle for survival. But the rate at which people now alter natural habitat has accelerated exponentially, and humanity's prized "skill"—its development and use of technology—is becoming a serious threat to our survival.

Direct Alteration of Habitat

We are altering our habitat most directly by intruding into the natural world—by expanding our cities into outlying natural places; allowing industrial-scale agriculture, lumber, power-generating, and mining operations to exploit natural areas; or, in countries with large numbers of landless poor people, forcing veritable armies of land-hungry peasants to seek a better life by converting natural areas to farms.

No country is immune to the problem of natural habitat alteration. In the first millennium, the native vegetation of Spain and much of the rest of Mediterranean Europe was effaced when goat herding became a major occupation. Goats also may have contributed to the desertification of the Sahara region in Africa—an area once covered with lush vegetation. European colonists and their descendants, armed with hatchets and saws, swept across North America in the nineteenth and early twentieth centuries, felling most of the continent's forests. Countries whose land is still in its natural state are now losing it rapidly, with the greatest losses occurring in countries with the greatest amount of wilderness.

Loss of Habitat in Figures Figures from the 1980s show that Africa south of the Sahara has lost an average of 65 percent of its original wildlife habitat, and tropical Asia has lost an average of 67 percent. Deforestation rates are alarming as well. Africa's worst case is the Ivory Coast, where 6.5 percent of the country's remaining forest disappears every year. Tropical America's most serious deforesters are Paraguay, which loses 4.7 percent of its forest cover every year, and Costa Rica, which cuts its forest at the rate of 4 percent per year. In tropical Asia, Nepal's forest is disappearing at a rate of 4.3 percent per year, with Sri Lanka in second place at 3.5 percent per year. (All of this information was originally from FAO, IUCN, and UNEP, and was cited in McNeely et al. 1990, 45–47.) The boreal forests of northern Canada and Siberia had long been ignored by the lumber industry because they were relatively inaccessible and forests in the Pacific Northwest of the United States were easier to cut. Now they, too, are suffering massive clear-cuts.

Especially Endangered Ecosystems Although most natural ecosystems on earth are threatened to some extent, certain places and certain types of habitats are especially vulnerable. Conservation International (see Organizations) has identified 17 "global biodiversity hotspots" that together occupy 2 percent of the earth's surface but harbor half of its species. These areas are notable not only for their high level of biodiversity but also for the extreme threats facing their delicate ecology. (See the Statistics, Illustrations, and Documents chapter of this book for a list of these hotspots.)

Especially at risk are forests, wetlands, rivers, islands, and lakes. These are of special concern for three reasons: because they are disappearing most quickly, their disappearance will have serious ripple effects, and they are home to a high proportion of endemic species.

Forests are threatened by logging, on one hand, and a common perception by landowners that their forested land would yield a greater profit if used for agriculture. Public decision makers are slowly learning that the other organisms living in forests, such as medicinal plants or rare cousins of our staple crops, could prove to be of greater economic worth than the lumber or the land the forest stands on. But for countries in need of immediate economic resources—and most tropical countries are in this category—it is a great temptation to sell off lumber rights.

Citizens of countries like Brazil have another motivation to convert forests to farms. Brazil's land is concentrated in the hands of a small, wealthy class, and these landholders wield much power in the government. So instead of redistributing the country's arable land more equitably, politicians use the vast Amazon Basin as an escape valve. Government policies encourage Brazil's landless peasants to colonize the Amazon Basin and convert its forest into small farms and ranches.

In the United States, with its stringent environmental protection laws, 1.5 million acres of wetlands—fresh and saltwater marshes and swamps, potholes (large, water-filled depressions mostly in the Midwest), estuaries (intertidal areas where river meets sea), and mangroves—are being destroyed annually, with only 95 million acres of the original 200 million acres of wetlands remaining (Liptak 1991). Imagine what is happening in countries that are less protective of their environment!

Rivers, the arteries feeding wetlands, are threatened as well, for many of the same reasons that wetlands are endangered. The mini-ecosystems of riverbeds depend on water that is clean, clear, and flowing—characteristics that disappear when modern society abuses rivers. Rivers are contaminated with chemicals used in agriculture and industry; those that flow through farmland and deforested areas fill with sediment that blocks sunlight and kills aquatic organisms; and the largest rivers are converted into reservoirs by huge dams built to generate electricity, drain off water for drinking or irrigation, or prevent floods.

Island habitats are prized for their endemism, but their flora and fauna are especially vulnerable. On islands that were uninhabited for most of their history, animals never developed a fear of humans or mammalian predators, which are generally less common on islands, since fewer immigrated from mainlands. So many endemic island birds, especially the numerous flightless species, have been decimated since the arrival of humans and their pets on remote islands. The famous Mauritius dodo was one of these, hunted to extinction in 1681 by Dutchmen living in a penal colony on the island of Mauritius. The last 11 Stephen Island wrens, a bird endemic to New Zealand's island of the same name, were killed by a lighthouse keeper's pet cat in 1894 (Frankel and Soulé 1981, 20–22). On large landforms, predation rarely is such a danger to species. But island prey has nowhere to flee; it is trapped by the confines of the island and thus is more readily extinguished than continental fauna.

Lakes are similar to islands because of their high level of endemism and their isolation from like ecosystems. Human interference in lake ecosystems—intentional or not—has had negative consequences for aquatic biodiversity. The entire population of cichlid fish species in Lake Victoria is threatened by one predator, the Nile perch, introduced by the Ugandan government in the 1920s as a game fish (Wilson 1992, 112). A more recent and possibly more serious threat to the endemic cichlids is the deforestation on the banks of Lake Victoria. Erosion from treeless slopes has clouded and eutrophied the lake, making it harder for cichlids to find members of their own species to mate with. As a result, the 300 species are collapsing into a murky collection of sterile hybrids (Yoon 1997). The Desert Fishes Council, an organization striving to conserve aquatic ecosystems in the deserts of the southwest United States, focuses on tiny fish that have survived for between 10,000 and 20,000 years in small, salty pools and streams where the temperature can rise to more than 120 degrees Fahrenheit. These rare fish face a double threat. Water pollution fouls their habitat, and the lakes dry up when people drill wells to underground aquifers and pump out their water.

A Sample Progression from Habitat Destruction to Extinction Once a forest, a wetland, a grassland, or any other habitat has been altered in a way that renders it inhospitable to the species that live there, the outlook is gloomy. The progression of events that leads to extinction of species is difficult to reverse.

Imagine that the habitat being destroyed is a large section of tropical rain forest. A lumber company works its way through a patch of forest, cutting every tree it can, compacting and rutting the bare earth with the tractors it uses to take the trees out to the main road. Every tree that falls brings down monkeys, sloths, bird nests, rare orchids and other epiphytes, and moss- and fungi-blanketed branches that form the microhabitat for tree-dwelling amphibians like poison dart frogs. The list could go on and on. The understory and ground-level vegetation is destroyed as well, and when the soil is compacted, this once humus-rich medium, full of microscopic bacteria and fungi, becomes as hard and infertile as blacktop. After stripping the land of its vegetation, the lumber company moves on to another plot, and then another, until only one small island of rain forest may remain in what used to be a vast tract.

What happens to the species that manage to survive the initial devastation? The animals that are relatively mobile flee to the remaining patch of undisturbed forest. That patch, however, had its own inhabitants, who now face a huge immigration crisis. The tiny island of forest is overcrowded, and food and shelter grow scarce. Certain plant species are overgrazed by mammals; fruits and berries are all plucked from their stems. Some animals escape this prison. They either make it far enough to find another undisturbed area or die of exhaustion and hunger en route. The animals that stay are subject to crowding, scarcity of resources, and the intense competition resulting from this situation. These are sources of stress for any organism, and stress facilitates the spread of diseases that kill off large numbers of the animals. The few that are left are forced to interbreed because they are isolated from other populations of their species, and the genetic poverty that comes from inbreeding eventually does them in.

Besides the pressure from the high numbers of individuals taking refuge in a small patch of habitat, another problem frequently arises. The number of species occupying a tiny area increases as well. In naturally occurring evolution, species specialize to avoid competing with one another (that is, certain animals eat only certain kinds of food, certain plants flower to attract pollinators at only certain times of the year or day, etc.). But when species that have evolved in different ecosystems are cast together, conflicts are bound to arise. The result is winners and losers, the defeated dying off.

Ecologists Robert MacArthur and E. O. Wilson articulated the theory of island biogeography in their book of the same name (1967). It stated that the area of a habitat, as well as how far it lies from other similar habitats, determines the number of species that can live there. Originally, the theory was used by ecologists to predict how many species could live in island habitats and how communities of species may have evolved. But in the 30 years that have passed since the book was published, it has assumed a new importance. Biologists concerned with conservation now use the theory to hypothesize how many species may disappear, and at what rate, when their once vast habitat is reduced to a series of "islands" floating in seas of devastated wildlands.

Besides reducing a habitat's area, the conversion of a large habitat to isolated islands changes the habitat in an important way. The ratio of edge, or border region, to interior of each forest patch is much higher. Some animals flourish in such edges, but many, including most bird species, simply will not breed in small patches of forest. Migratory songbirds, for example, make up 80–90 percent of the species breeding in large North American forests, but less than half of those are found in small, isolated forest patches (DiSilvestro 1993, 197).

Migratory Species Animals that migrate may face multiple threats, because all of the habitats they depend upon—their summer and winter homes and the resting spots in between—may be subject to the same destruction. Caribou in Arctic Canada move from their summer grazing lands in the tundra ecosystem to coniferous forests farther south in the winter. Both of these areas are threatened: tundra by potential oil drilling and pipeline installation, the forests by the lumber industry. Migratory songbirds have been surveyed since 1948 and have shown a consistent decline in their population. Their summer habitats have been converted to suburbia, and their winter habitat in Central America is being transformed into plantations (DiSilvestro 1993, 197). North American raptor species also suffer from the loss of their Central American rest stops. European storks are threatened by overuse of chemicals in Europe, by Israeli fighter jets along their annual migration path, and by toxic garbage dumps at one stopover in Egypt. Sea turtles, which nest on the same beach every year but spend the rest of their lives in the ocean, depend not only on clean oceans free of plastic trash but also on the preservation of their nesting beaches. Unfortunately, many of these beaches have been discovered by the tourist industry. Bright hotel lights as well as human or animal activity on beaches at night dissuade turtles from laying their eggs onshore. The eggs that are successfully laid are often dug up or stepped on by people and dogs (Durrell 1986; *Sinking Earth* 1987).

Difficulties Specific to Plants Because they are not mobile like animals, plants cannot flee to an undisturbed refuge when their habitat is destroyed. The fertility of a few lone plants left after a clear-cut is contingent upon pollination, germination of seeds, and survival of young plants. But this is often a complex process. Many plants need animate pollinators such as insects, bats, birds, even small mammals. If these animals have fled to a refuge far away, their return to pollinate the plant is doubtful. The seeds of some plants just fall to the ground and germinate there, but other plants benefit from having their progeny grow farther away. Those seeds depend upon some sort of vehicle—attaching themselves to an animal's foot or fur and detaching at some later time elsewhere. Other seeds will germinate only if they have passed through the gut of a certain animal whose digestive juices wear away the hard coating of the seed and allow the embryo inside to germinate in the earth. Even if a seed successfully germinates, the struggle is not yet over. The young plant may require conditions such as shade and moisture that will not be present if it is growing in a deforested environment. It is important to note that in none of the mass extinctions of the past have plants become extinct at the rate they are disappearing currently.

Indirect Habitat Alteration

Not all the harm we do to other organisms on earth has to do with our direct destruction of individual habitats. Other more indirect effects result from our contamination of the atmosphere and the earth's surface, including its land and water.

Atmospheric Contamination Our widespread burning of fossil fuels and wood releases an inordinate amount of carbon dioxide into the atmosphere. Besides polluting the air, this excess carbon dioxide may be a cause in the climatic warming the planet is currently undergoing. Scientists have not yet determined the full effect of climatic warming, but they warn of several possibilities. There may be changes in hydrological cycles, with altered weather patterns. Very moist areas may become drier, and vice versa. Since plants have evolved to cope with the amount of moisture in their habitat, drastic weather changes could cause extinction of some species and variations in composition of ecological communities. Humid forests adapted to abundant rainfall may die if the area dries up. Global warming will result in warmer climates around the world. A warmer climate is more conducive than a cold one to microbes, some of which cause plagues and diseases. Scientists expect invasions of these maladies as the climate heats up. A warmer climate will also force animals adapted to cool temperatures to migrate to higher elevations or poleward. However, adapting quickly enough to the environmental conditions of new ecosystems may not be possible for most animals. But the most devastating effect of all might be due to changes in the soil's chemical makeup. The bacteria and fungi that aid plants in absorbing mineral nutrients from the soil have evolved in accordance with certain temperature or moisture conditions. We do not know how sensitive they are to variations. If they cease to function, the productivity of the land will plummet and famine will spread (Alders 1994, 9).

Industrial production of such chemicals as CFC coolants generates by-products that perforate the ozone layer. In addition to accelerating global warming, the main effects detected so far have been higher incidences of skin and eye disease in animals. How the loss of the shield will affect photosynthesizing plants and other life-forms is not yet known. Other atmospheric contaminants emitted by industry, such as organic compounds containing chlorine, are taken up by terrestrial or aquatic plants. The estrogen-mimicking effects of these compounds, which are suspected to cause cancer, intensify as the compounds concentrate in higher levels on each step up the food web (Freeman 1996).

Water Contamination Aquatic areas are particularly vulnerable to contamination. Rain rinses off whatever is on the land: pesticides and fertilizers from farms, lawns, and golf courses, and petroleum by-products from city streets. The toxic runoff drains into rivers, lakes, wetlands, and oceans. Factories pump chemical waste into streams and rivers. These toxins concentrate where the water flow slows—primarily in wetlands and lakes. Lakes have been declared "dead" after their fish die or they are asphyxiated by eutrophication—excessive growth of algae and depletion of oxygen due to high fertilizer content. Because fish, mollusks, and amphibians hatch and grow up in wetlands, and juveniles are much more vulnerable to toxins than adults, these classes of animals are particularly endangered by water pollution.

Other Causes of Extinction

Hunting has been a well-publicized cause of the extinction of some animals. The extinction of the Mauritius dodo and the

Stephen Island wren, discussed above, were accomplished by human and feline hunters, facilitated by the defenselessness of the prey. North America has been the site of other tragic and now famous extinctions by hunting. Passenger pigeons, once so abundant that flocks darkened the skies as they flew across the continent, were extinguished in less than half a century by bird hunters. The last one died in 1910. American bison were similarly targeted by hunters in the 1800s and were even shot for sport from trains soon after railroads were built. The only reason the bison were not extinguished was that the Bronx Zoo in New York saved a handful of them, bred them successfully, and replenished herds in protected areas (Cohen 1995).

Exotic invaders, whether intentionally or unintentionally introduced, can crowd out native species. One famous case is the zebra mussel, which was introduced into the Great Lakes ecosystem when it was pumped out of a European ship's ballast. It has been extremely successful in its colonization and has forced out native mollusk species. The eucalyptus tree, valued by foresters for its perfectly straight trunks and rapid growth, has been introduced from its native Australia to almost every country in the world. This tree sends deep roots into the ground and soaks up so much moisture that other plant species cannot grow.

Naturally occurring biodiversity is not all that is at risk. Traditional food crops are also disappearing. The thousands of species that have traditionally served as food for humankind, however, have been reduced to a handful of basic foodstuffs in countries where indigenous agricultural practices have been abandoned. Corn, wheat, rice, and potatoes are now the four most important carbohydrate sources for most people on earth. These foods have literally invaded the diets of many nations. One writer recalls signs shouting "EAT WHITE BREAD" in Nigeria. That country, along with the rest of Africa and many Latin American countries as well, was pushed to import wheat in the 1950s and 1960s, a time when powerful nations like the United States had huge wheat surpluses. African farmers no longer had a local market for their traditional drought-resistant crops, such as sorghum, finger millet, pearl millet, tef, fonio, and African rice, so they stopped growing them (Glantz 1996).

The United States, too, has suffered a tremendous agricultural erosion. The genetic base for the meat, vegetables, fruits, grains, and legumes we consume has been steadily diminishing as agriculture becomes more mechanized and small-scale diversified farming yields to large plots of a single crop (monocropping). The more than 7,000 varieties of apples and 2,300 varieties of pears grown in the United States a century ago have been reduced by at least 85 percent. Today, 90 percent of the eggs purchased in the United States are laid by one breed of hen, the white leghorn (Teitel 1992, 4). This genetic erosion renders our food base extremely vulnerable to pests and diseases.

An additional danger to biodiversity is caused by our heavy dependence on chemical pesticides. Pesticides are usually designed to protect one particular crop but often cause incalculable damage to other organisms. As rain rinses sprayed crops, toxic runoff poisons waterways, wetlands, and aquatic organisms within them. And in some cases, crop pollinators are harmed, too. Honeybees pollinate many crops, but the use of insecticides containing methyl parathion has killed about 30 percent of the bees in some areas of the country (Baird 1997).

Indian biodiversity activist Vandana Shiva (see Biographical Sketches) argues that traditional subsistence farming, as practiced in Indian villages and many other communities around the world, deserves as much respect and protection as naturally occurring biodiversity. She contends that the trend toward industrial-scale monocropping, with varieties developed by seed companies that select for high productivity over all other qualities, rob farmers, and the rest of the world as well, of invaluable biological resources.

Biodiversity is threatened by humankind on virtually every front: through human civilization's expansion into wilderness, industrial emissions that alter the global climate, an onslaught of toxic chemicals on our waterways, and the intentional erosion of the genetic pool of our food supply. We are bombarded with information about the loss of the riches in earth's natural treasure chest. So why isn't conservation of biodiversity an international priority? Many people accept the justification that economic survival is only possible through the plundering of our natural resources. The dichotomy is framed in headlines like this one: "In Suriname's Rain Forests, a Fight over Trees vs. Jobs" (DePalma 1995) or in the head-to-head conflicts between environmentalists and lumbermen such as that occurring in the old-growth forests of the northwestern United States.

Overcoming Threats to Biodiversity

Costa Rican conservationist Rodrigo Gámez once observed that "the North American forest disappeared because what was valued was land and wood. The rest had no value" (Wallace 1992,

151). When diverse biological resources are not valued, destruction of wilderness results. Huge natural areas continue to disappear from the earth every day, cut by profit-craving corporations, gouged by hungry peasants, or smothered by urban and industrial development. At the same time, indirect threats to biodiversity are more insidious: climatic change triggered by atmospheric contamination, water pollution, and intentional reduction of agricultural diversity. Given what is now common knowledge about the importance of conserving the diversity of genes, species, and ecological communities, how can the destruction be halted?

Worldwide Strategies

There are nearly as many strategies, programs, and hard-working individuals currently addressing the biodiversity crisis as there are natural areas to preserve. International organizations have issued biodiversity conservation directives to guide individual nations as they plan their own conservation strategies. One of the most comprehensive of these international documents is the Convention on Biodiversity, approved at the 1992 Earth Summit in Brazil (see the Statistics, Illustrations, and Documents chapter). Government commissions have designed responses to their own biodiversity conservation challenges, and nonprofit organizations, driven by thousands of devoted individuals, work with governments and at the grassroots level.

Economic Impediments to Conservation

On the global scale, conservation of biodiversity is challenged by elements of the world's current economic system. The scale used to measure a nation's wealth is based on how much a country "produces" for export. A productive country sells something to other countries, either raw goods (such as tree trunks, minerals mined from the earth, or an agricultural product in an unprocessed form) or manufactured or processed goods (such as sawed planks or wheat flour) made from its own raw materials or raw materials from another country. Traditional economic ratings do not include the value of intact ecosystems and their resources, the functions they perform to maintain global health, or the compounds that will eventually be discovered there and will prove to be of great economic worth. Vandana Shiva points out that a country where most inhabitants sustainably harvest products from the forest for their own use would be considered poor. This assessment is shortsighted and discourages sustainable use of biodiversity.

Aggravating the situation is the worldwide practice of offering government subsidies for so-called productive activities such as agriculture, fishing, grazing, logging, and mining. Recent reports estimate that these subsidies—amounting to $500 billion worldwide—actually result in environmental damage costing the world $500 to $900 billion annually. The subsidies not only encourage many environmentally destructive activities, they also contribute to the overuse of such expensive inputs as pesticides, fertilizers, and water (Crossette 1997).

Tropical countries contain most of the world's species in their biodiverse ecosystems. But as rich as they are biologically, they are financially impoverished. Most of these countries owe huge sums of money to international lending institutions as a result of intense borrowing during the past few decades. Unfortunately, the economic situations in most debtor countries didn't improve with massive borrowing—in many cases economic situations actually got worse. Now these governments are being forced by their creditors to implement austerity plans that further pauperize their poorest citizens. Poverty tempts individuals to violate protected natural areas (often the only uninhabited land left) and to cut timber or to eke out a living by raising a few head of cattle or subsistence crops. Indebted governments, or corrupt officials, are tempted to earn quick cash by selling off lumber or mining rights to large operations. The companies usually pay much less than the resources are worth on the world market, but debtor governments, in their desperation, feel forced to accept any cash offer.

When negotiating with tropical countries, world economic powers would do well to view these nations as custodians of an unexplored treasure, a bank of biological resources of value to the entire world, a bank to which the world must contribute.

Population Pressure

Another major threat to biodiversity is population pressure. There is a bitter debate between so-called Malthusians, who feel that population must cease growing at its current elevated rate or people will soon exhaust the world's resources, and those who feel that there would be enough for everyone if the resources were not concentrated in the hands of the wealthiest class. A fairer distribution of economic resources would certainly relieve the poverty of huge numbers of people and would reduce rampant overconsumption in wealthy countries. But there can be no doubt that the exponential growth of the human race and its need for food and shelter has led to an alarming decline of biological resources. If projections for continued population growth are correct, the outlook for the survival of the earth's biodiversity is bleak. But many local, national, and international actions are being taken to conserve biodiversity.

Managing Wildlands to Conserve Biodiversity

North American ecologist Daniel Janzen believes that potentially unlimited treasures are buried in the rich biodiversity of places such as his adopted country, Costa Rica. Janzen has devoted much of his energy not only to deciphering the intricacies of tropical ecology but also to establishing ambitious conservation programs in Costa Rica that are becoming models for the rest of the world. He outlines three steps for conserving biodiversity: "1. Save it. 2. Know what it is. 3. Use it sustainably" (Janzen 1992, 28). These steps are deceptively simple, especially in the most biodiverse but economically strapped countries, where meeting the basic needs of growing populations requires most of the attention and funds of the government.

Conservation in Publicly Owned Wilderness

Best exemplified by the national park system in the United States, "saving it" was once accomplished by declaring a beautiful geographic feature a national park, marking its boundaries, assigning park guards to keep out poachers, and expecting that this would result in the preservation of the ecological communities and species dwelling within. This strategy is no longer viable, given increasing pressure on what little unpopulated land remains and recent biological findings that many species need huge areas of interconnected habitats.

National park managers and planners are learning that many of the species they want to protect in existing or future national parks may migrate to ecosystems outside park borders during the year seeking warm, cool, rainy, or dry weather. They may mate or bear young far from where they spend their adult life. They may require a vast range for hunting or grazing. Knowing this—and taking into consideration that global warming may force even more extensive migrations—wildlife managers have realized they must either create national parks much larger than yet economically or politically feasible or ensure the conservation of important ecosystems on adjacent land, whether in public or private hands.

Janzen, working with other Costa Rican conservationists, designed a response to this problem. Guanacaste National Park is an annex to an older national park encompassing most of the habitats that the original park's species depend on at some point in their lives. Local subsistence farmers and ranchers were included in the development of the idea and are now employed as park guards and research assistants. The park service's earlier policy of staffing the park only with personnel from outside the area had alienated locals and led them to feel that the park was for tourists from the country's capital and abroad. Their resentment was heightened by a ban on hunting on lands that had been open to them before the area was declared a park, since hunting was a traditional practice that *campesinos*—peasant farmers as they are known in Spanish-speaking countries—undertook both for sport and to supplement their families' diet. Janzen and his colleagues have been able to convert farmers who just a decade before felt animosity toward a park system into strong allies of conservation. Although poachers still cause occasional problems, Guanacaste National Park is considered a model for wildlands managers throughout the developing world.

Africa's unique conservation programs have also advanced beyond the traditional U.S. parks model. Communal Areas Management Programme for Indigenous Resources (CAMPFIRE) is the name of a well-known, successful project in Zimbabwe. The country is divided into local districts, many of which contain park land inhabited by the type of wildlife that safari hunters and ecotourists come to see. The parks department, which in the past had jurisdiction over the land, granted authority over the wildlife to about half of the country's local districts. Along with this authority have come hunting quotas corresponding to the species' growth rates, so that hunting can remain a sustainable activity. Wealthy foreign safari hunters spend about $1,000 per day on lodging, food, and a guide, in addition to paying sizable trophy fees for the animals they kill. This income, along with what is generated by other types of tourism, is divided up by the local district's villages and can be used in any way the local council wishes. Conserving wildlife has great costs for local people: crop destruction, loss of livestock, and sometimes human deaths. But with an arrangement like CAMPFIRE, villages are receiving compensation and feel a greater commitment to protecting their local wild animals.

Maasai Mara National Park in Kenya is the most popular spot in East Africa for watching lions, elephants, rhinos, giraffes, and other spectacular species. It is surrounded by grazing ranches belonging to the Maasai people, whose traditional livelihood is shepherding livestock. The Maasai shepherds, like the park neighbors in Zimbabwe, pay a price to live so close to wildlands: the loss of their animals to wild predators like lions and to diseases spread by wildlife. They are prohibited from hunting in the national park but can kill wild animals that stray onto their ranches. The government has been experimenting with ways to convince the Maasai to collaborate in the protection of the national park's habitat and species. Cash contributions to community funds from national park entrance fees seem to be the most successful so far (Bonner 1993; Lewis and Carter 1993).

The Costa Rican and African parks, as well as numerous others throughout the world, are paid for by the citizens of these countries, by tourist entrance fees, and by hefty donations from such international conservation organizations as the World Wildlife Fund, the Nature Conservancy, and the World Conservation Union. None of the foreign sources are wholly dependable, however. International organizations like to build up a project and then cut it loose to live off its own efforts. Tourism, too, can wax and wane with the political climate, economic conditions in tourists' home countries, and other unpredictable idiosyncrasies of the trendy travel industry.

Conservation on Private Lands

Conservationists know that preservation of biodiversity cannot be limited to lands owned by the government and are increasingly encouraging private owners to retain natural areas on their land. Costa Rica helps neighbors of its 20-some national parks, refuges, and reserves to develop small-scale businesses that cater to park visitors. The park service there also encourages neighbors with undeveloped land to consider these areas nature reserves and invite paid guests to visit. Park managers sometimes promote visitation to these private reserves over already overvisited parks, both to relieve pressure on public land and to ensure that conservation is profitable for locals. The result has been an explosion of private reserves, now organized into their own association, which collaborate closely on conservation issues with the government's park service. In areas where a national park does not include all of the necessary habitats for its species, these private reserves prove very important for wildlife.

Another option for conservation of biodiversity on private land is to motivate farmers and ranchers to work in the least environmentally damaging way possible. In tropical countries, grassroots organizations promote the cultivation of "analog forests," which mimic natural forests and were widely used in ancient times by Mayan people and others to grow their food. They include a variety of tree, bush, and vine species so that there are several layers of vegetation in which photosynthesis occurs. Each species is chosen for what it can contribute both to the collection of foodstuffs produced and to the cultivation system itself. Some promote the recycling of nutrients; others shade plants that cannot tolerate full sun. Ground-level crops help the soil retain moisture and nutrients (Knight 1995). In the United States, farms and ranches occupy half of the country's land area. The Nature Conservancy assists conservation-minded farmers and ranchers in the regeneration of ecologically rich portions of their land. The U.S. Department of Agriculture's Conservation Reserve Program, which in past decades paid farmers to let fields lie fallow to reduce crop surpluses, now pays farmers to allow less productive yet ecologically valuable sections of their farms to return to a natural state.

The World Conservation Union (IUCN) is active throughout the world, with local teams promoting hundreds of efforts that improve local economies through grassroots environmental conservation projects. It carries out several wildlife management projects in Central America. One of them helps *campesinos* from the lakeside village of El Jocotal, El Salvador, conserve nesting habitat for a wild duck species. *Campesinos* benefit from a continuous supply of eggs to supplement their diet, and the ducks are lovingly protected by villagers. Another project benefits Guatemalan villagers who live within the Mayan Biosphere Reserve. They are taught techniques for managing wild animal populations so that animals are hunted at a rate that will not deplete the population (Solís and Cruz 1996).

Sustainable Use of Resources

Projects like those conducted by the IUCN promote sustainable development, which can be defined as "the use of natural resources without destroying the possibility for future generations of using the resources in the same way, if they wanted" (Sandlund et al. 1992, 9). Widespread adoption of sustainable practices would result in conservation of biodiversity not only in public parks and reserves but on private lands as well.

Sustainability in Indigenous Lifestyles

Sustainable development may be a new concept for national governments and nongovernmental conservation organizations, but it is an ancient way of life for most traditional cultures. Some of the practices that indigenous peoples have evolved over thousands of years of living in what we now consider wilderness can serve as examples for cultures striving to develop lifestyles that are not destructive to nature.

Indigenous peoples living in tropical rain forests, for example, find everything they need in the forest: food, medicine, hunting equipment, construction materials, and clothing. They know how to harvest all of these goods without depleting them. However, they usually do not manufacture products from them that can generate export income. Perceiving that these indigenous peoples do not make any "productive" contribution to the national economy, governments often seek to relocate them and turn over their land to more productive tenants. Many groups of indigenous people are routed from their homelands by non-indigenous, government-supported colonists, or lumber, mining, or agriculture companies. While countless indigenous groups have succumbed to destruction of their habitat, either dying off or becoming assimilated into the dominant nonindigenous culture (one-third of Brazil's 270 indigenous groups have disappeared during this century [A. Wallace 1994, 86]), a few have been able to defend their homes and lifestyles.

The indigenous inhabitants of the KéköLdi Indigenous Reserve of Costa Rica's Atlantic lowlands have succeeded in maintaining their lifestyle through a very intentional survival strategy. They have marked the borders of their reserve in order to prevent incursions by squatters. Despite their familiarity with the legal and political structures of the outside world, the KéköLdi people follow a traditional lifestyle guided by religious beliefs requiring them to use forest resources at a sustainable rate. They have even gone one step further by establishing an iguana regeneration project on their land. They raise these nutritious reptiles for their own food supply and to repopulate their reserve with them. The KéköLdi people have described their way of life in their book *Taking Care of Sibö's Gifts* (see the Print Resources chapter).

The Xishuangbanna Dai people of the Yunnan province of southwestern China are another indigenous group with beliefs and practices that revere biodiversity. They consider the forest the cradle of humanity because water originates in forests, land is fed by water, and food is grown in land fed by water. Without forests there would be no food and without food, no people. Near each community they protect "Holy Hills," forest-covered hills where the gods are said to reside with the flora and fauna that are their sacred companions. No hunting, wood chopping, or cultivation is allowed there. The inhabited areas, too, are biodiverse systems and include home gardens with a variety of crops, different types of agroforestry, and temple yards with holy plants (native species said to be Buddha's favorites). Their care of temple yard gardens has preserved such rare and valuable plant species as the gingko, which otherwise may have become extinct (Shengji 1993).

Throughout the world there are thousands of indigenous peoples such as the BriBri and Cabécar people living on the KéköLdi Reserve in Costa Rica and the Xishuangbanna Dai in China, whose age-old lifestyles have a lot to teach those trying to conserve biological diversity. The delegates writing the 1992 Convention on Biodiversity understood the contribution indigenous people make to understanding and conserving nature, and linked their continued existence to the survival of biodiversity.

Sustainability Resulting from More Complete Knowledge of Biodiversity

For indigenous cultures, sustainable use of resources is dictated by religious teachings and sheer necessity, and is made possible by an intimate familiarity with the workings of their habitat. For nonindigenous peoples, Daniel Janzen's second step in the biodiversity conservation process, "Know what it is," can facilitate his third step, "Use it sustainably." Detailed knowledge about biological resources, coupled with sophisticated management and business savvy, can convert biodiversity into a large-scale economic resource.

With much effort by Janzen and other Costa Rican conservationists, Costa Rica has developed what is now a model for other countries, its National Institute for Biodiversity (INBio). The institute opened its doors in 1989 with a mission (see the Statistics, Illustrations, and Documents chapter) to carry out an inventory of all living organisms in Costa Rica. It has enlisted a corps of parataxonomists, residents of rural areas who receive training in identification of the taxa that INBio focuses on (primarily insects and plants) and collect species in the field. They label their specimens with the time and place they were found, perform an initial classification, and turn in their collections to headquarters, where technicians further classify them. Curators keep track of collections after the technician's work is done, and professional taxonomists regularly visit to help with unidentified specimens. INBio uses a computerized system so that eventually the information—which species occur in what locations throughout the country—

will be easily accessible to the public. This information is gaining more importance with the surge of international interest in biological resources, and one of the motives for creating INBio was to develop a mechanism by which the country could derive an economic benefit from its biodiversity.

The biodiversity of a country without strong national institutions such as INBio can be exploited without compensation. Costa Ricans retain a painful memory of the British Technology Group's (BTG) discovery of the powerful natural nematocide—or parasitic nematode killer—extracted from the *Lonchocarpus costaricaensis* tree. BTG has made millions of dollars from the discovery; Costa Rica has earned nothing. But the worst example of this phenomenon occurred in Madagascar, where a flower named the rosy periwinkle was discovered by chance. Two alkaloids from this species, vinblastine and vincristine, are used to treat childhood leukemia and Hodgkin's disease. Their manufacture and sale yield a profit of $180 million a year to the drug companies. None of these profits have been returned to Madagascar, where lack of conservation funding has deforested 95 percent of the country (Wilson 1992, 283).

INBio was established in part to prevent a reoccurrence of this experience. INBio now collaborates with international companies interested in exploring the chemical compounds of Costa Rica's biodiversity and has signed agreements with the pharmaceutical firm Merck (see the Statistics, Illustrations, and Documents chapter) and the perfumery Givaudan-Roure. The agreements give the companies access to Costa Rica's biodiverse forests while assuring that a share of the profits from the exploitation of any chemical compound discovered there will return to Costa Rica to help pay for conservation projects.

Other countries, especially those with rich, unexplored biodiversity, have been inspired to develop their own biodiversity institutions, with the support of INBio. So far Mexico, Kenya, Panama, Colombia, the Philippines, and Suriname have set up their own biodiversity inventory or data management projects (A. Lovejoy 1997).

Sometimes private corporations initiate their own experiments in sustainable use of biological resources. Shaman Pharmaceuticals of San Francisco is banking on the knowledge of indigenous shamans in its exploration for useful medical compounds. Amazon medicine men use more than 6,000 plants to treat their peoples' ailments (A. Wallace 1994, 86), and only a small percentage of these have been analyzed by Western laboratories.

As compensation to the communities whose traditional knowledge it depends upon, Shaman donates a portion of each expedition's budget to a public health project initiated by the tribe visited. Projects have ranged from installing a potable water system to paying for visits from a dentist. A physician in each team shares Western cures for medical problems that the local shaman has not been able to heal. Shaman Pharmaceuticals has developed two drugs that will soon enter the market: an oral medication for respiratory viruses and a topical agent for herpes. If the company makes a profit from these medications, a portion of its earnings will be returned to the communities where the drugs originated.

Conserving Individual Endangered Species

Conservationists agree that the best approach to preserving biodiversity is through the protection of natural habitat. By protecting an entire ecological community—the larger the better—all of the species within will have a better chance of survival. And if local residents can become partners in the effort, changing their ways if necessary to live more harmoniously with their wild neighbors, everyone involved will benefit. Unfortunately, more drastic, individualized approaches have become necessary to conserve highly endangered species. This direct intervention takes several forms.

Treaties and Laws for Endangered Species Protection

One of the main motives for poachers of endangered species is the high price they or their body parts command on the international market. A meeting held in 1973 about commercial trade in endangered species resulted in an agreement called The Convention on International Trade in Endangered Species of Wild Flora and Fauna (CITES). CITES, which has been signed by 120 countries, regulates international trade in endangered species by maintaining three lists of species for which trade is regulated. Appendix I includes species most in danger of extinction. Commercial trade in these species is not allowed; they can be traded only between zoos and captive breeding projects. Appendix II lists species that could become endangered without regulation; permits can be issued for their trade. Appendix III includes species that are considered endangered in their home country. CITES is enforced by customs or fish and wildlife agents of each country. Although CITES could be a potent tool for controlling poaching, underfunding of inspection teams and high-level government corruption are recurring problems that limit its effectiveness. International trade in endangered species still garners $5 billion per year.

Most CITES signatories have enacted their own system of recognizing and protecting endangered species. In 1973, the U.S. Congress passed the Endangered Species Act, which requires the Fish and Wildlife Service (USFWS—see Directory of Organizations) to maintain a list of species that are endangered (facing extinction) or threatened (likely to become endangered soon). The USFWS is charged with restoring populations of listed species so that they can eventually be removed from the list. Since protecting an endangered species often necessitates the protection of entire ecosystems, more than one dam or suburban development has been halted by the USFWS because construction would have endangered or extinguished a species. Numerous conservation organizations have assumed a watchdog role over the USFWS to assure that it complies with its mandate.

Periodically the IUCN publishes the *Red Data Books*, which list by country or continent all the species in danger of extinction (see the Statistics, Illustrations, and Documents chapter). Because these books contain authoritative lists compiled by scientists active in each area, they exert a powerful influence. A new listing, or a focus on a particular area, can command the immediate attention of national governments and international conservation organizations.

Ex-situ Conservation Approaches

If a species identified as endangered is "charismatic" enough—in other words, if it has broad appeal—scientists may intervene with an *ex-situ* conservation project. These are labor-intensive approaches to saving a single species and are usually undertaken only when the *in-situ* approach—conservation of the species through conservation of its habitat—seems impossible. Sometimes poachers render a species' natural habitat too dangerous. Or there may be so few individuals left that reproduction is unlikely. The *ex-situ* method involves taking members of an endangered species out of the wild (or using individuals previously captured or domesticated) and breeding them in captivity to conserve their genetic diversity and replenish their dwindling numbers. In the best of cases, individuals born in captivity can be reintroduced into the wild. In situations where the species' natural habitat is gone or they remain under threat by poachers, the hope is that the species and its genetic diversity can be maintained until reintroduction becomes possible. *Ex-situ* success stories include replenishment and reintroduction of the American bison, Arabian oryx, and the Brazilian tamarin—all species saved from a very close brush with extinction.

Ex-situ conservation projects are located in the same countries as the natural habitat of the endangered species, and at botanic gardens, zoos, and aquaria throughout the world. Two international networks that link *ex-situ* projects, the Species Survival Plan (SSP) developed by the American Association of Zoological Parks and Aquariums, and the International Species Inventory System (ISIS), allow for the exchange of information on species being propagated in captivity and can help arrange for genetically desirable couplings between animals housed at different institutions. Breeders act as genetic managers, assuring the survival of equal numbers of males and females, preventing any family to become much larger than the others, and preventing family members to interbreed (Foose 1986, 150). In cases of extremely endangered species, scientists have experimented with implanting test tube embryos in surrogate mothers of related species. An endangered Grant's zebra embryo, for example, was successfully implanted in a common horse in 1984 (Dresser 1988, 299).

Botanical gardens and seed and germ plasm banks operate as *ex-situ* conservation centers for endangered plant species. Two of the most important centers in the United States are the New York Botanical Garden and the Center for Plant Conservation at the Missouri Botanical Garden (see Organizations).

The *ex-situ* approach is seen by most conservation biologists as a last resort. Its expense and labor intensity can draw away resources from *in-situ* conservation efforts that conserve both individual species and the ecosystems in which they live. Furthermore, domestically raised individuals of some species (including plants) may never be successfully reintroduced into the wild. Left in their natural habitat, plants and animals have a chance to continue evolving to meet ever-changing conditions. But by the time a species is released into its original habitat, environmental conditions may have changed so that it will have to deal with a situation its ancestors never faced in the wild.

Halting the Indirect Alteration of Habitat

Just as there are international agreements like CITES to protect particular species, there are accords to halt the emissions of contaminants affecting air and water quality. The 1987 Montreal

Protocol on the Protection of the Ozone Layer required signatory nations to reduce their production of ozone-depleting chlorofluorcarbons (CFCs). The Convention on Climate Change, signed at the 1992 Earth Summit, and the 1997 Kyoto Summit treaty underlined and proposed solutions to the problem. But obstacles remain to the reduction of the so-called greenhouse gases. These gases, which include carbon dioxide, methane, nitrous oxide, hydrofluorocarbons, perfluorocarbons, and sulfur hexafluoride, are thought to lead to a greenhouse effect, in which heat is trapped and the earth warms.

Costa Rica is experimenting with a plan to counteract excess atmospheric carbon dioxide by conserving designated carbon dioxide sinks. Clients (public utilities, industries, and even national governments) pay the Costa Rican government or private owners to maintain a sink of a certain capacity for an agreed-upon period of time. There are still many details to be worked out, such as which agency will regulate the seller's compliance with the agreement (that is, not cutting down the trees) and what the motivation for paying for carbon sinks will be in addition to the owners' environmental conscience. Some scientists also question whether tropical forests are the best possible absorbers of carbon dioxide. If the remaining questions can be resolved, this experiment could become a model for how wealthy countries that produce the most carbon dioxide can support conservation in poorer countries (Liddell and Escofet 1997).

The Ramsar Convention on Wetlands of International Importance Especially as Waterfowl Habitat of 1971 binds the 54 signatory nations to protect wetlands, including rivers, lakes, swamps, coastal areas, tundra, floodplains, and shallow ocean areas (McNeely et al. 1990, 138). In response to this convention, many countries have implemented their own wetlands acts. But despite international agreements and national laws, the wetland contamination problems described in previous sections persist.

Restoration of Damaged Ecosystems

When human civilization has all but eradicated a native ecosystem, ecological restoration may be the last resort. The tallgrass savanna restoration project in the environs of Chicago has brought back more than 67,000 acres of an ecosystem that had not only disappeared decades ago but had also been completely forgotten even by prairie ecologists. Miraculously, tallgrass savanna animals—presumed to be extinct because they hadn't been seen in recent history—returned when their habitat was restored (A. Wallace 1994, 96–105; Budiansky 1995, 233–236; Stevens 1995). The arduous work of helping nature regenerate is being undertaken all over the world with small-scale successes that inspire similar efforts. Proponents of restoration point out that not only does this work benefit the inhabitants of restored ecosystems, it also allows the people doing the work to become part of an ecological process.

Conclusion

A tremendous amount of work has been dedicated to preserving biodiversity. International bodies such as the United Nations Environment Program (UNEP) have organized worldwide summits like the 1992 meeting in Rio de Janeiro, which called international attention to biological degradation. National governments have enacted stringent environmental laws restricting pollution and have mandated the protection of biodiversity on public and private lands. Thousands of nongovernmental organizations pour expertise and funds into conservation strategies and projects. But long-term conservation of biodiversity depends upon a drastic change in how most humans view nature. Indigenous people living in ways that sustain natural habitats are a small minority in most parts of the world. Greater numbers of people live in urban or suburban environments too far from functioning ecosystems to develop an understanding of their dependence upon nature and of how nature works.

Social ecologist Stephen Kellert studies how people view nature and has discovered an alarming ignorance. People have prevalent and strong phobias about the life-forms most important to our survival. Invertebrates—90 to 95 percent of all the species and 90 to 95 percent of biomass (or dry weight) on earth—are the life-forms that humans most despise. Their grotesque appearance, their insidious ability to penetrate our living spaces, the size of their population, the rapidity of their reproduction, and an ancient collective memory of their potential damage to people and food may be to blame. Kellert's study, in which people were interviewed about their feelings toward invertebrates, notes that a majority "expressed willingness to eliminate whole classes of animals altogether, including mosquitoes, cockroaches, fleas, moths, and spiders." This ignorance of the roles animals play in cleaning up the earth is dangerous; 99 percent of all fecal excretions on the planet are decomposed by invertebrates (Stevens 1993).

It should be possible for people to cultivate a different frame of mind, one that knows, respects, and reveres nature. E. O. Wilson, preeminent expert on biodiversity, has developed a hypothesis known as biophilia (an affection for life-forms), which holds that people have an innate fascination with other life-forms. He hopes that accessing our own biophilia will help us change the way we treat the world. (See Print Resources chapter for Wilson's *Biophilia*, the original articulation of the thesis, and *The Biophilia Hypothesis*—a compilation of responses to the thesis by other scientists, edited by Wilson with Stephen Kellert.)

Saving the earth, a desire motivated by biophilia, is a cause that has gained popularity recently, with proponents recommending all sorts of actions—from boycotting hamburger chains to recycling soft drink cans to driving stakes into hardwood trees targeted by loggers. But we probably do not need to worry about saving the earth itself: during its 5 billion years of existence it has been subject to bombardment by meteors and comets, to huge variations of temperature leading to the formation of glaciers and the melting of polar ice caps, to collisions between continents, to tremendous uplifts of land, to earthquakes, tidal waves, and volcanic activity. The earth will continue evolving as its molten center moves under its tectonic plates. We cannot do anything about these geological events or the extraterrestrial influences on our planet.

But preserving the biosphere, the thin envelope of life that the planet supports, is something we can do. We can respond to direct and indirect habitat alteration by lobbying governments to conserve and work toward sustainability. We can support the nonprofit organizations that encourage sustainable economic activities all over the world, and especially in biodiverse areas. One author offers these suggestions for individual actions: have no more than two children, which is a "replacement rate" (one child per parent, which does not increase the overall population); study the environment for a better understanding of its beauty and complexity; think about the consequences of your profession and lifestyle; respect the planet's resources; join both a local and a national conservation organization; use your vote to influence environmental policy (Durrell 1986, 215).

An expert on consumer power advises that shoppers have the power to stem genetic erosion of our foodstuffs. By choosing organic produce, shoppers can encourage farmers to quit using pesticides. They can also show farmers that they support greater genetic diversity by buying unusual varieties and species of fruits and vegetables. On the other hand, purchasing fruits and vegetables out of season is not recommended because they are grown in tropical countries or in the Southern Hemisphere. Since the American market is so lucrative, farmers in these places stop growing their own native crops and switch to the apples, peaches, and grapes that they can export to the United States (Teitel 1992). Eating less meat is another indirect contribution to natural habitat conservation. EarthSave (see Organizations) reports that more than two-thirds of the grain produced in the United States is fed to livestock; 1 pound of U.S. feedlot beef requires 2,500 gallons of water, 12 pounds of grain, and 35 pounds of topsoil.

Although these personal actions are not on the same scale as international treaties or national environmental laws, they do allow individuals to act as partners in what needs to be a planetwide quest to preserve biodiversity.

Many natural habitats have been inalterably degraded. Thousands of species have already disappeared. Uncountable numbers of traditional indigenous people have abandoned their sustainable lifestyles. That is past tense, and there is little we can do about these losses besides study what once existed and learn not to repeat mistakes. But there is still much to save and there are many ways to help.

References

Alders, Hans. 1994. "Towards Biodiversity in Politics." In *Biodiversity and Global Change*, edited by O. T. Solbrig, H. M. van Emden, and P. G. W. J. van Oordt, 9–12. Oxon, UK: CAB International.

Associated Press. 1996. "Tree Variety Sets Record." *New York Times*. 12 November: B6.

Baird, R. E. 1997. "Suspects Emerge in Bees' Demise." *Colorado Daily*. 22 April: 4, 7.

Blake, Beatrice, and Anne Becher. 1997. *The New Key to Costa Rica*. Berkeley, CA: Ulysses Press.

Bonner, Raymond. 1993. *At the Hand of Man: Peril and Hope for Africa's Wildlife*. New York: Alfred A. Knopf.

Budiansky, Stephen. 1995. *Nature's Keepers: The New Science of Nature Management*. New York: Free Press.

Cohen, Daniel. 1995. *The Modern Ark: Saving Endangered Species.* New York: G. P. Putnam's Sons.

Crossette, Barbara. 1997. "Subsidies Hurt Environment, Critics Say before Talks." *New York Times.* 23 June: A3.

DePalma, Anthony. 1995. "In Suriname's Rain Forests, a Fight over Trees vs. Jobs." *New York Times.* 30 September: 1, 4.

DiSilvestro, Roger L. 1993. *Reclaiming the Last Wild Places.* New York: John Wiley & Sons.

Dresser, Betsy L. 1988. "Cryobiology, Embryo Transfer, and Artificial Insemination in *Ex-Situ* Animal Conservation Programs." In *Biodiversity,* edited by E. O. Wilson, 296–308. Washington, DC: National Academy Press.

Durrell, Lee. 1986. *State of the Ark: An Atlas of Conservation in Action.* New York: Doubleday.

Ehrlich, Paul, and Anne Ehrlich. 1981. *Extinction: The Causes and Consequences of the Disappearance of Species.* New York: Random House.

Erickson, Jon. 1991. *Dying Planet: The Extinction of Species.* Blue Ridge Summit, PA: TAB Books.

Erwin, Terry L. 1988. "The Tropical Forest Canopy: The Heart of Biotic Diversity." In *Biodiversity,* edited by E. O. Wilson, 123–129. Washington, DC: National Academy Press.

Foose, Thomas J. 1986. "Riders of the Last Ark." In *The Last Extinction,* edited by Les Kaufman and Kenneth Mallory, 141–165. Cambridge, MA: MIT Press.

Frankel, O. H., and Michael E. Soulé. 1981. *Conservation and Evolution.* Cambridge, UK: Cambridge University Press.

Freeman, Karen. 1996. "Nuclear Byproduct Cleanses Stubborn Smokestack Pollutants." *New York Times.* 3 December: B8.

Glantz, Michael. 1996. "Africa Finds 'Lost' Crops." *Boulder Daily Camera.* 29 July: C1.

Gould, Stephen Jay. 1993. *Eight Little Piggies: Reflections in Natural History.* New York: W. W. Norton.

Hopkins, John. n.d. "Preserving Native Biodiversity." San Francisco, CA: Sierra Club Public Affairs.

Iltis, Hugh. 1988. "Serendipity in the Exploration of Biodiversity." In *Biodiversity,* edited by E. O. Wilson, 98–105. Washington, DC: National Academy Press.

Janzen, Daniel. 1992. "A South-North Perspective on Science in the Management, Use, and Economic Development of Biodiversity." In *Conservation of Biodiversity for Sustainable Development,* edited by O. T.

Sandlund, K. Hindar, and A. H. D. Brown, 27–52. Oslo, Norway: Scandinavian University Press.

Knight, Katherine. 1995. "Farm in the Forest." *Tico Times*. 3 November: 1, 8.

Lewis, Dale, and Nick Carter, eds. 1993. *Voices from Africa: Local Perspectives on Conservation*. Washington, DC: World Wildlife Fund.

Liddell, Jamie, and Guillermo Escofet. 1997. "'Carbon Bonds': Doubts Abound." *Tico Times*. 16 May: 1, 11.

Liptak, Karen. 1991. *Saving Our Wetlands and Their Wildlife*. New York: Franklin Watts.

Lovejoy, Annie. 1996–1997. Series of personal communications. (Lovejoy is coordinator of public relations at INBio, Santo Domingo de Heredia, Costa Rica.)

Lovejoy, T. E. 1995. "Quantification of Biodiversity." In *Biodiversity: Measurement and Estimation,* edited by D. L. Hawksworth, 81–87. London: Chapman & Hall.

Lugo, Ariel. 1988. "Estimating Reductions in the Diversity of Tropical Forest Species." In *Biodiversity*, edited by E. O. Wilson, 58–70. Washington, DC: National Academy Press.

MacArthur, Robert, and E. O. Wilson. 1967. *The Theory of Island Biogeography*. Princeton, NJ: Princeton University Press.

Mittermeier, Russell A. 1988. "Primate Diversity and the Tropical Forest." In *Biodiversity*, edited by E. O. Wilson, 145–154. Washington, DC: National Academy Press.

Pimm, Stuart L. 1991. *The Balance of Nature? Ecological Issues in the Conservation of Species and Communities*. Chicago: University of Chicago Press.

Plotkin, Mark J. 1988. "The Outlook for New Agricultural and Industrial Products from the Tropics." In *Biodiversity*, edited by E. O. Wilson, 106–116. Washington, DC: National Academy Press.

Prance, Ghillian T. 1986. "The Amazon: Paradise Lost?" In *The Last Extinction*, edited by Les Kaufman and Kenneth Mallory, 62–106. Cambridge, MA: MIT Press.

Raup, David M. 1988. "Diversity Crises in the Geological Past." In *Biodiversity*, edited by E. O. Wilson, 36–50. Washington, DC: National Academy Press.

———. 1991. *Extinction, Bad Genes or Bad Luck?* New York: W. W. Norton.

Ray, G. Carleton. 1988. "Ecological Diversity in Coastal Zones and Oceans." In *Biodiversity*, edited by E. O. Wilson, 36–50. Washington, DC: National Academy Press.

Sandlund, O. T., K. Hindar, and A. H. D. Brown, eds. 1992. *Conservation of Biodiversity for Sustainable Development.* Oslo, Norway: Scandinavian University Press.

Shengji, Pei. 1993. "Managing for Biological Diversity Conservation in Temple Yards and Holy Hills: The Traditional Practices of the Xishuangbanna Dai Community, Southwest China." In *Ethics, Religion and Biodiversity: Relations between Conservation and Cultural Values,* edited by Lawrence S. Hamilton, 118–132. Cambridge, UK: White Horse Press.

Sinking Earth. 1987. One video in a five-part series entitled *Only One Earth,* Julian Pettifer, host. A BBC-UNESCO production, distributed by Public Media Education, Chicago, IL.

Solís Rivera, Vivienne, and Juan Carlos Cruz Barrientos. 1996. *Experiencias de Manejo de Vida Silvestre en Centroamérica (Wildlife Management Experiences in Central America).* San José, Costa Rica: IUCN.

Stevens, William K. 1993. "Bugs Keep Planet Livable Yet Get No Respect." *New York Times.* 21 December: C1, C16.

———. 1995. *Miracle under the Oaks.* New York: Pocket Books.

———. 1997. "New Suspect in Ancient Extinctions: Disease." *New York Times.* 29 April: B7, B10.

Survival in Nature. 1990. One part of a two-part episode from the video series entitled *Frontiers in Science and Nature.* Produced by Scientific American, distributed by SVE & Churchill, Chicago, IL.

Terborgh, John. 1992. *Diversity and the Tropical Rain Forest.* New York: Scientific American Library.

Wallace, Aubrey. 1994. *Green Means: Living Gently on the Planet.* San Francisco, CA: KQED.

Wallace, David Rains. 1992. *The Quetzal and the Macaw: The Story of Costa Rica's National Parks.* San Francisco, CA: Sierra Club Books.

Wilson, E. O. 1988. "The Current State of Biological Diversity." In *Biodiversity,* edited by E. O. Wilson, 3–18. Washington, DC: National Academy Press.

———. 1992. *The Diversity of Life.* Cambridge, MA: Belknap Press of Harvard University Press.

Yoon, Carol Kaesuk. 1996a. "Lake Victoria's Lightning-Fast Origin of Species." *New York Times.* 27 August: B5–B6.

———. 1996b. "Parallel Plots in Classic of Evolution." *New York Times.* 12 November: B5, B8.

———. 1997. "Rainbow of Cichlid Fish Colors Is Disappearing." *New York Times.* 23 September.

Chronology 2

This chapter charts several chronologies. First, there is the decrease in biodiversity, marked by habitat destruction and the extinction of species. Then there is the development of a body of knowledge about biodiversity, estimates of its extent, new discoveries of the intricacies of interrelationships between species in ecological communities, and evidence that biodiversity is far more important to our continued existence than most people previously believed. A third component of this chronology is the public response to the growth of scientific knowledge about biodiversity and its loss: conservation measures taken by governments, nonprofit organizations, and the grassroots—local citizens trying to protect their own piece of biodiversity.

Researchers interested in the history of the U.S. conservation movement may want to consult DiSilvestro (see reference list); many of the government conservation-related events were taken from the first chapter of his book. The sources for information on other events are cited to help researchers locate good primary source material.

1681 On the Indian Ocean island of Mauritius, the flightless dodo is

1681 *cont.* hunted to extinction by European colonists. Because the island bird previously had no predators, it has no fear of humans and is an easy target.

1741 Steller's sea cow, a huge marine mammal second in size only to whales, is discovered by Western scientists in the Bering Sea's shallow coastal waters off Alaska.

1768 The meaty, defenseless Steller's sea cow (see 1741), easy prey for hunters, becomes extinct.

1871 Congress creates the Commission on Fish and Fisheries to investigate the declining catch in U.S. waters.

1886 The Department of Agriculture's Division of Economic Ornithology and Mammalogy is set up and assigned by Congress to study interactions of agriculture, birds, and mammals.

1891 The New York Botanical Garden is founded and sends its botanical researchers to the West Indies, Florida, and the western states to gather specimens of plants heretofore unknown to science.

1892 The Sierra Club is founded by naturalist John Muir. Most of its early members and projects are centered in the West, especially in California.

1896 William Brewster, one of the foremost ornithologists of his time, founds the Massachusetts Audubon Society for the protection of birds.

1900 The Lacey Act illegalizes the importation of foreign wildlife without a permit and the transport over state lines of wildlife killed in violation of state laws. But at this point there are few state laws against poaching.

1903 President Theodore Roosevelt establishes the first protected wildlife refuge on Pelican Island off Florida's Atlantic coast, the first of more than 50 refuges he created during his presidency. During the twentieth century, the National Refuge Systems grows to include more than 400 refuges.

1905 The U.S. Forest Service is established to administer the cutting of timber in national forests.

The Audubon Societies established around the turn of the century join to become a national group, which works on conservation legislation and owns and administers private wildlife sanctuaries.

1911 This is the last year that wolves are seen in Newfoundland, Canada. Twenty years later, as a result of U.S. government bounties for killing wolves, wolves grow rare in the United States.

1913 Congress passes the Migratory Bird Act protecting game species like ducks and geese and migrating songbirds and plume birds.

1914 The last passenger pigeon, Martha, dies in the Cincinnati Zoo. In the early 1800s, the species had been so numerous that flocks numbered into the billions. Hunting so reduced their population that the last known nesting site was recorded in 1878, 36 years before Martha died (Stewart 1978, 14).

1916 The Migratory Bird Treaty Act between the United States and Canada is signed, requiring those two countries to collaborate on protective measures for migrating birds (Stewart 1978, 18).

The National Park Service is created to manage existing national parks.

1929 The Migratory Bird Act of 1913 (see above) was extended to authorize the federal government to buy land necessary for the conservation of migratory birds.

1934 The Migratory Bird Hunting Stamp Act required waterfowl hunters to purchase a federal stamp for their state hunting licenses. The proceeds were spent on creating wildfowl refuges.

1935 The Wilderness Society is formed to assist in the protection of public lands.

1936 The United States and Mexico sign a migratory bird act similar to the one that the United States and Canada had signed 20 years earlier. It provides habitat for migratory birds throughout the entire North American continent (Stewart 1978, 18).

Hunters concerned with conservation form the National Wildlife Federation.

1949 *A Sand County Almanac and Sketches Here and There,* by Aldo Leopold, is published posthumously. It includes the oft-quoted affirmation that "A thing is right when it tends to preserve the integrity, stability, and beauty of the biotic community. It is wrong when it tends otherwise."

1950 The fish and mammal/bird agencies established in 1871 and 1886, respectively (see above), merge to become the Fish and Wildlife Service of the Department of the Interior. It manages the extensive National Wildlife Refuge System.

1958 The National Seed Storage Laboratory is established in Fort Collins, Colorado. Its purpose is to preserve seeds of as many as possible of the valuable plants native to the United States. The seeds are stored at a temperature of –196 degrees centigrade and should last into eternity. Ninety-nine percent of the seeds will suffer no damage, and scientists estimate their life span at 100 billion years. Whether the plants could survive in that future environment, without having evolved to adapt to it, is questionable, however (Nabhan 1989).

1960 The first section of the Arctic National Wildlife Refuge is signed into existence. It doubles in size when another act is signed 20 years later. This vast, pristine refuge is the habitat for caribou, snow geese, tundra swans, wolves, moose, and grizzly bears, along with about 7,000 native people. It is the only place in the United States where polar bears mate. Most of these animals spend summer months in the Arctic National Wildlife Refuge and winter further south in Canada. The refuge is threatened by the petroleum industry, which drills nearby and constantly seeks permission to drill within the refuge (DiSilvestro 1990, 179).

1962 Botanists Hugh Iltis and Don Ugent set out on an expedition to collect wild tomatoes in Peru. A tomato geneticist at the University of California–Davis, Dr. Charles Rick, soon breeds the genes from one wild variety into a commercial variety and develops a fruit with an unusually high sugar content. This improved sweetness is estimated to be worth $8 million annually to the tomato industry. The original research trip, which yielded 1,000 collections that were surrendered to geneticists, cost $21,000 (Iltis 1988, 102–103).

Rachel Carson's *Silent Spring* is published, sounding an alarm at the overuse of synthetic pesticides, which damage many species besides insect pests.

1963 The Eli Lilly pharmaceutical company places on the market an antileukemia drug extracted from the rosy periwinkle, a flower endemic to Madagascar. By 1985, worldwide sales of this drug have reached $100 million, 88 percent of which is profit. None of the money reaches Madagascar, however, a desperately poor country whose rich biodiversity is being destroyed faster than anywhere else on earth.

Due to the use of the pesticides dieldrin and DDT, the peregrine falcon population in Britain has been reduced by 44 percent since World War II—only 350 pairs survive. Once use of these pesticides ceases in the late 1960s, the population increases until it reaches numbers greater than at any time during this century (Cade 1988, 282).

1967 Robert MacArthur and E. O. Wilson publish *The Theory of Island Biogeography* (Princeton, NJ: Princeton University Press, 1967), which puts forward the theory that the area of an island and its distance from the mainland affect the number of species that can live on it. A correlating theory is that as habitat is reduced, extinctions will occur at a predictable rate. This is the theoretical base for most estimates of how

1967 *cont.* many species will become extinct as habitats are eliminated by humans expanding into nature.

1968 James Lovelock first describes the Gaia Hypothesis, which proposes that the biosphere is a self-regulating mechanism—producing warming carbon dioxide when the atmosphere is too cool, and producing cooling oxygen when the atmosphere is too warm.

1969 The Desert Fishes Council is formed by a group of scientists concerned about disappearing aquatic habitats in the southwestern United States.

1970 Eighty percent of the corn sown this year in the United States contains the same gene, T. cytoplasm, which farmers like for the corn's high yield and rapid growth, although it renders the variety vulnerable to the blight *H. maydis*. In previous years, when the gene was not so common, *H. maydis* had never proven a major problem. But when so much of the corn planted contained this gene, the blight was able to tear through monocropped cornfields, destroying 15 percent of that year's crop (Shiva 1991).

The United Nations Conference on the Environment, held this year in Stockholm, Sweden, brings issues of the environment into the international spotlight for the first time.

1971 The Convention on Wetlands of International Importance Especially as Waterfowl Habitat is signed at a meeting of nations in Ramsar, Iran. This is the only convention covering a particular type of habitat, and it defines wetlands as rivers, lakes, swamps, coastal areas, tundras, floodplains, and ocean areas less than six meters deep at low tide. Each of the more than 50 signatory nations is required to designate at least one site to be included in the List of Wetlands of International Importance.

1972 The California sea otter is protected by state law, and conservationists are able to restore the reduced otter population to its original size. The otters resume their predation of sea urchins and clams. Sea urchins no longer mow offshore kelp at such a furious rate, and the kelp beds attain their previous luxuriance, once again providing shelter and food to numerous species of ocean fish.

The United States bans the pesticide DDT, a synthetic pesticide that caused cancer in humans and whose overuse had been decried in Rachel Carson's *Silent Spring* ten years earlier.

The United Nations Educational, Scientific, and Cultural Organization (UNESCO) convokes a meeting in Paris that results in the Convention Concerning the Protection of the World Cultural and Natural Heritage. The more than 100 signatory nations agree to protect sites of cultural and natural importance. The convention calls for the establishment of World Heritage Sites, chosen from the most biodiverse areas of the world.

1973 The U.S. Congress enacts the Endangered Species Act, which follows two previous, weaker versions in the 1960s. The Endangered Species Act lists endangered or threatened species. The first category includes those species in danger of extinction in their entire range, and the latter includes species likely to become endangered soon. The act prohibits U.S. citizens from hunting the listed species or destroying their habitat. The U.S. Fish and Wildlife Service and the National Marine Fisheries Service are charged with listing new species and designing recovery plans.

The International Species Inventory System (ISIS) is set up to help prevent inbreeding of endangered animals in captivity. Members of ISIS include 200 zoos and captive breeding projects worldwide. They contribute genealogical information for their animals, and that information is used to make genetically ideal matches between animals (Dresser 1988, 299).

1974 The women of Reni, a village in northern India, angry that a commercial logging operation plans to deforest their watershed, respond by wrapping their arms around the trunks of trees as the lumber company

1974 *cont.*

workers approach. The Chipko or "tree hugging" movement prevented the deforestation, and eventually the government officially protected the watershed. Since that time, the Chipko movement has transformed itself into an extensive reforestation effort, spreading out into other villages in the region.

The Nature Conservancy sets up the first State Natural Heritage Inventory in South Carolina. Its job is to gather and manage data on biodiversity and organize it so that it can be useful for conservation efforts. Within ten years there is a Natural Heritage Inventory in nearly every state and similar institutions in several Latin American and Caribbean countries.

The International Board for Plant Genetic Resources (IBPGR) is established by the United Nations' Food and Agriculture Organization (FAO) to combat genetic erosion in the world's food supply. The IBPGR gathers genetic materials for more than 50 crops and stores them in more than 100 gene banks (Williams 1988, 241).

1975

According to an FAO study, 15 million acres of arable land in North and Central America have been covered by cement and asphalt since 1945.

The Convention on International Trade in Endangered Species (CITES) comes into force to regulate international trade in species that are considered endangered or threatened in their countries of origin. As of 1996, CITES has been signed and ratified by 120 countries.

1977

Steve Packard midwives the birth of the North Branch Prairie Project, which over the next two decades evolves into a giant effort to restore the forgotten tallgrass savanna of the Midwest. Over the years 5,000 volunteers have worked to restore 67,000 acres of tallgrass savanna.

Dr. Wangari Maathai launches the Green Belt Movement in Kenya, designed to provide renewable fuelwood resources for subsistence farmers, supplement the income of poor rural women by paying them to plant and care for trees, prevent soil erosion and desertification, and conserve natural habitat.

A wild perennial species of corn, *Zea diploperennis*, is discovered by a Mexican college student on a 10,000-foot mountain in Jalisco, Mexico. It is proven resistant to the seven most serious diseases that affect corn, and economists calculate that this species, if used to create a perennial hybrid, could be worth almost $7 billion to agriculture. Ten years later, the Mexican government and UNESCO declare a 350,000-acre Biosphere Reserve that protects the habitat of *Zea diploperennis* and many other species (Iltis 1988; Shiva 1991).

1979 British ecologist Norman Myers publishes the first findings on tropical rain forest destruction. He estimates that 1 percent of the entire area of tropical rain forests disappears each year, and with it, one-quarter of 1 percent of the world's species. This revelation alarms many in the scientific community, notably E. O. Wilson, and inspires a sense of responsibility to conserve the natural ecosystems that support their profession.

A meeting in Bonn, Germany, produces the Convention on Conservation of Migratory Species of Wild Animals. This convention obligates participating parties to protect endangered migratory species and provides a framework for international agreements (McNeely et al. 1990, 139).

Entomologist E. O. Wilson invents the term *biophilia*, which he defines as "an inborn affinity human beings have for other forms of life, an affiliation evoked by pleasure, or a sense of security, or awe, or even fascination blended with revulsion" (Wilson 1994, 360).

1980 Australia protects 11,800 square kilometers of coral reefs, islands, and surrounding ocean as the Great Barrier Reef Marine Park. Four hundred species of coral and 1,500 fish species inhabit this area (Myers 1984, 99).

1980 *cont.* Climate change reports reveal that atmospheric carbon dioxide levels have increased by 30 percent since 1850. Carbon dioxide in the atmosphere prevents heat from radiating into space, which results in a warmer climate (the greenhouse effect). These same reports predict that the level of carbon dioxide could increase by an additional 75 percent before the year 2060 (Myers 1984, 116).

The World Conservation Strategy (WCS), a document cowritten by five major world organizations (IUCN, WWF, UNEP, FAO, and UNESCO), with contributions from 400 scientists, proposes an integrated, global approach to environmental problems. The WCS urges countries to devise their own National Conservation Strategies (NCS) addressing their most critical problems. By 1990, 40 countries have initiated their own NCS (Myers 1984, 168; McNeely et al. 1990, 56).

The Committee on Research Priorities in Tropical Biology determines that five times more taxonomists than there are currently will be needed to classify the biodiversity thought to exist. As future estimates of species roll in, the lack of trained taxonomists becomes more and more problematic (McNeely et al. 1990, 72).

1981 A wild population of 100 to 120 black-footed ferrets is discovered in Wyoming. Within one year their population is reduced by 90 percent because their prey, prairie dogs, die off from the plague and the ferrets themselves contract canine distemper. A technically complicated and very expensive federal program attempts to restore their population (Seal 1988, 291).

1982 The National Cancer Institute finds that Americans have a 31 percent chance of contracting cancer before they turn 74. The major source of the cancer is environmental pollution (Myers 1984, 122).

Entomologist Terry L. Erwin of the Smithsonian Institution publishes a new, drastically larger estimate of the number of species in the world based on his research on host-specific beetles in a Panamanian rain forest. Whereas previously there were thought to be a total of about 1.5 million species in the world, Erwin estimates that there are 30 million species of insects alone (Erwin 1988, 123).

The United Nations passes the World Charter for Nature, which includes the statement that "all species and habitats should be safeguarded to the extent that it is technically, economically, and politically feasible."

1984 The Center for Plant Conservation is founded. It is a network of botanical gardens, seed banks, and arboreta committed to preserving the endangered plants of their regions. Within a decade, 25 major institutions join, together preserving more than 500 endangered species as seeds, germ plasm, and living plants.

1985 Zebra mussels, an exotic freshwater species of clam-like mollusks, are introduced accidentally into the Great Lakes. The larval form was probably carried from Europe in the ballast tanks of a ship and pumped out upon arrival. The zebra mussel has no predators in the Great Lakes ecosystem, and the native clams have no defenses against it. It reproduces abundantly and within a decade has migrated all the way down the Mississippi River to New Orleans.

A five-year campaign has turned the golden lion tamarin, endemic to the degraded Atlantic Coast forests of Brazil, into a national conservation symbol. Of the 21 species of monkeys in this region, 80 percent are endemic and 67 percent are endangered. Thanks to an enthusiastic publicity program that put the tamarin onto the nation's phone books and postage stamps, and into parades and theater productions, the tamarin and its habitat are better understood by the Brazilian people and may even be recovering (Mittermeier 1988, 148–149).

1985–1986 A watershed two years for publicizing the threats to biodiversity. The scientific and conservation organizations that sponsored major international meetings

1985–1986 *cont.* on the subject include the Smithsonian Institution (December 1985), World Wildlife Fund (September 1986), National Academy of Science and the Smithsonian Institution (September 1986), and the New York Zoological Society (October 1986).

1987 More biological research rolls in proving that tropical rain forests harbor more species than any scientist expected. E. O. Wilson identifies 43 species of ants (the same number found in the entire landmass of the British Isles) in the Tambopata Reserve in Peru; Peter Ashton finds 700 species of trees in ten one-hectare plots in Borneo (as many as in the whole North American continent).

1989 The Costa Rican National Biodiversity Institute (INBio) is founded (see Organizations).

The Thai government bans logging after 100 years of actively encouraging it. The ban is won by a coalition of students, scientists, environmentalists, traditional farmers, development workers, and the media. Local grassroots groups, supported by these allies, also win battles against several hydroelectric dams, which impact biodiversity through deforestation, local climate change, loss of soil fertility, colonization, and fishery depletion (Lohmann 1991).

The Council for the Preservation of Rural England reveals that the last 40 years have seen a drastic change in the face of the English countryside. Each year 4,000 small farms are taken over by large companies. Since 1949, 25 percent of hedgerows, or bushes and trees bordering fields, have been torn out. At least 50 percent of heaths, meadows, and wetlands have been destroyed, with the ensuing loss of all of the wild plants and animals that inhabited them. More than 300 plant species are officially endangered (Myers 1984, 153).

1990 Lisa Conte raises $3.8 million to found Shaman Pharmaceuticals, a California drug company that takes a shortcut to developing drugs. Instead of testing hundreds of compounds randomly for active compounds as traditional pharmaceutical companies do, Shaman is guided by the knowledge of traditional medicine men, its research teams testing medicinal plants already in use. (See Organizations.)

1992 The United Nations Conference on the Environment and Development (UNCED, commonly referred to as the Earth Summit) is held in Rio de Janeiro, Brazil, resulting in several international agreements: the Rio Declaration on Environment and Development, a list of 27 short principles that should guide nations' environmental actions; Agenda 21, a 500-page document to be used as a guide for achieving sustainable development in the twenty-first century; the Forest Principles, which address national sovereignty issues in forest conservation; the Convention on Climate Change, which addresses emissions of greenhouse gases; and the Convention on Biodiversity, which offers guidelines for the conservation of biodiversity and the utilization of its products.

1993–1994 Between 80 and 96 of the 200 to 400 Siberian tigers remaining in Siberian forests are killed by poachers during this period of chaos and corruption in the dismantled Soviet Union. Biologists have not determined whether the number of tigers left will be sufficient to maintain the species. If not, Siberian tigers will join four other species of tiger that have met extinction during the last 50 years: the Bali tiger, extinguished in the 1940s; the Caspian/Iranian tiger, extinguished in the 1970s; the Java tiger, extinguished in the 1980s; and the Chinese tiger, most likely extinct in the wild due to the demand in Chinese medicine for its ground-up bones. The only tiger species still viable is the Bengal tiger, of which between 3,000 to 5,000 are said to remain.

1994 Rain forest canopy biologists who ascend to the treetops in contraptions ranging from dirigibles to electric gondolas to cranes to rope-and-pulley systems meet at the Marie Selby Botanical Gardens in Sarasota, Florida, to share their findings. Among the anecdotes from this new perspective: the Australians have discovered a tree-mite mutualism. The tree's

1994 *cont.*	leaves form houselike shelters that protect mite colonies from wind and rain, and in return, the mites attack any invading, harmful fungus.
1996	A team of researchers studying the cichlid fishes of Africa's Lake Victoria discovers that the 300 species there have evolved in less than 12,000 years, at what evolutionary biologist Amy McCune called "rates of speciation that have not even been imagined" (Yoon 1996).
	Federal officials loosen the locks of the Glen Canyon Dam 15 miles above the Grand Canyon in Arizona in order to create seasonal flood conditions that will accommodate the plants and animals adapted to spring and summer floods. Since the dam was built in 1963, putting an end to the floods, there has been an ecological degradation that has caused near extinction for several species endemic to the Grand Canyon. Floods help those species by stirring up nutrient-rich sediment and pushing out nonnative invader species. The officials must continue to experiment with water volume and duration of the floods in order to come up with the conditions most similar to those of the previous floods (Stevens 1997a).
	The domestic honeybee colonies of Boulder, Colorado, have declined by one-third in one year. Two culprits are suspected: an exotic parasitic mite from Europe, which was first spotted in the area in 1987, and the widespread use of Penncap-M synthetic pesticide, which contaminates the pollen that bees take back to their hives. The pesticide is preserved during the winter in the pollen, and when that pollen is fed to larvae in the spring, it poisons them. Whole generations are lost in affected hives. Boulder County Apiary Inspector Tom Theobald predicts that many residents will see their fruit trees and gardens become infertile. This phenomenon is not limited to Boulder; it is happening all over the world (Baird 1997).
1997	Mammalogist Ross MacPhee of the American Museum of Natural History in New York proposes

an addition to the prevailing two-part theory on the major extinction of 13,000 years ago that claimed mammals such as saber-toothed cats, mastodons, huge beavers, and stag moose. He claims that along with a changing climate and more effective human hunters, diseases carried by human immigrants to the American continent helped to extinguish these mammals (Stevens 1997b).

Tropical conservationists identify a main cause for the decrease in the population of migratory songbirds. Their winter homes in Central America are being converted into monocropped coffee plantations. In past years, coffee growers have interplanted coffee bushes with many other species of shade and fruit trees. But a new breed of "sun" coffee does not tolerate shade, so plantation owners are cutting down the other trees. As a result, songbirds such as orioles, wood thrushes, and warblers are left homeless half of the year, and many do not survive the hardship (Witze 1997).

Two plant ecologists, Martin Cipollini and Doug Levey, discover that the bitter alkaloids in some fruits and vegetables have evolved in order to dissuade predators from eating them. However, they don't seem to affect the "right" disperser—the one the plant depends upon to disperse its seeds. With further research, alkaloids that don't taste bad may be discovered. These would be of great value to the produce industry, which seeks a method to prevent produce from rotting without altering its taste (Yoon 1997).

As the U.S. Congress prepares to reevaluate the 1973 Endangered Species Act, the Congressional Research Services releases its calculations that in the two and a half decades since the ESA's enactment, 1,520 species have been listed as endangered or threatened, 11 of them have been recovered from endangerment, and 7 have become extinct. Several years earlier, the General Accounting Office found that 80 percent of endangered and threatened species continued to decline in population (Campbell and Spivak 1997).

1997 *cont.* An unprecedented 31-page special report in *Science* journal, written by 20 top environmental scientists, warns that people have seriously altered the biosphere. Among the damage done: half of the earth's surface water is consumed before it reaches the oceans—this causes lethal siltation and salinization; between 40 and 50 percent of the planet's land surface has been altered by people, destroying natural habitat, causing hotter temperatures and lower rainfall in some places, eroding soil, and polluting the water; the amount of nitrogen being cycled has more than doubled (due to activities such as mining, fertilizer manufacturing, and fossil fuels burning), which results in water pollution, fish kills, and eutrophication; one fourth of all bird species, one-sixth of all mammals, one twelfth of all plants, and one-twentieth of all fish species are either extinct or on the verge of extinction. Says coauthor Jane Lubchenco, "We are tinkering with our life support systems...[but] we don't realize the consequences of what we are doing" (Dewar 1997).

References

Baird, R. E. 1997. "Suspects Emerge in Bees' Demise." Colorado Daily. 22 April: 4, 7.

Campbell, Matt, and Jeffrey Spivak. 1997. "Endangered Species Act Marked by Progress, Loss." Denver Post. 27 July: 35A.

Dewar, Heather. 1997. "Earth's Life-Support Systems Rated Seriously Ill." Denver Post. 25 July: 1A, 8A.

DiSilvestro, Roger L. 1990. Fight for Survival: A Companion to the Audubon Television Specials. New York: John Wiley & Sons.

Dresser, Betsy L. 1988. "Cryobiology, Embryo Transfer, and Artificial Insemination in *Ex-Situ* Animal Conservation Programs." In Biodiversity, edited by E. O. Wilson, 296–308. Washington DC: National Academy Press.

Erwin, Terry. 1988. "The Tropical Forest Canopy." In Biodiversity, edited by E. O. Wilson, 123–129. Washington DC: National Academy Press.

Iltis, Hugh. 1988. "Serendipity in the Exploration of Biodiversity." In Biodiversity, edited by E. O. Wilson, 98–105. Washington, DC: National Academy Press.

Leopold, Aldo. 1949. A Sand County Almanac and Sketches Here and There. New York: Oxford University Press.

Lohmann, Larry. 1991. "Who Defends Biological Diversity? Conservation Strategies and the Case of Thailand." In Biodiversity: Social and Ecological Perspectives, by Vandana Shiva et al., 77–104. London and Penang, Malaysia: Zed Books Ltd. and World Rainforest Movement.

McNeely, Jeffrey A., et al. 1990. Conserving the World's Biological Diversity. Gland, Switzerland: IUCN; and Washington, DC: WRI, CI, WWF-US, and the World Bank.

Mittermeier, Russell A. 1988. "Primate Diversity and the Tropical Forest." In Biodiversity, edited by E. O. Wilson, 145–154. Washington, DC: National Academy Press.

Myers, Norman. 1984. Gaia: An Atlas of Planet Management. New York: Doubleday.

Nabhan, Gary. 1989. Enduring Seeds: Native American Agriculture and Wild Plant Conservation. San Francisco, CA: North Point Press.

Seal, Ulysses S. 1988. "Intensive Technology in the Care of Ex Situ Populations of Vanishing Species." In Biodiversity, edited by E. O. Wilson, 289–295. Washington, DC: National Academy Press.

Shiva, Vandana. 1991. "Biodiversity, Biotechnology and Profits." In Biodiversity: Social and Ecological Perspectives, by Vandana Shiva et al., 43–58. London and Penang, Malaysia: Zed Books Ltd. and World Rainforest Movement.

Stevens, William K. 1997a. "A Dam Open, Grand Canyon Roars Again." New York Times. 25 February: B7, B12.

———. 1997b. "New Suspect in Ancient Extinctions: Disease." New York Times. 29 April: B7.

Stewart, Darryl. 1978. From the Edge of Extinction. New York: Methuen.

Williams, J. Trevor. 1988. "Identifying and Protecting the Origins of Our Food Plants." In Biodiversity, edited by E. O. Wilson, 240–247. Washington, DC: National Academy Press.

Wilson, Edward O. 1994. The Naturalist. Washington, DC: Island Press.

Witze, Alexandra. 1997. "Not Quite All Sing Coffee's Praises." Denver Post. 22 May: 1E, 4E.

Yoon, Carol Kaesuk. 1994. "Biologists Master the Real Scene of Forest Action." New York Times. 22 November: B7, B11.

———. 1996. "Lake Victoria's Lightning-Fast Origin of Species." New York Times. 27 August: B5–B6.

———. 1997. "Wild Fruits' Flavor, Toxicity Serve Good Purpose." Denver Post. 18 May: 41A.

Biographical Sketches

So many people are hard at work on biodiversity issues that it would be impossible to introduce more than a handful in a chapter as brief as this one. Scientists, conservationists, writers, philosophers, economists, gardeners, planners, and indigenous people trying to maintain their traditional lifestyle are among the countless individuals dedicating themselves to conserving biodiversity. This chapter is not meant to be comprehensive in its coverage but rather seeks to represent the variety of individuals who are actively working to preserve biodiversity.

For more biographies of environmentalists worldwide, consult the list of recipients of the Goldman Environmental Prize. This annual prize is bestowed on six grassroots environmental activists representing six world regions: North America, South/Central America, Asia, Island Nations, Africa, and Europe. The Goldman Environmental Prize maintains a homepage with lists and descriptions of past and current winners (http://www.goldmanprize.org/goldman/) and publishes a booklet with mini-biographies of winners. The organization can be contacted at One Lombard Street, #303, San Francisco, CA 94111 (415-788-9090; fax 415-788-7890; e-mail gef@igc.apc.org).

Additional useful resources for biographies include *Green Means* and *Environmental Heroes: Success Stories of People at Work for the Earth*, both of which profile environmentalists and their projects (see Print Resources chapter for more complete descriptions).

Edwin Bustillos (b. 1965)

This Mexican agricultural engineer was honored with a Goldman Environmental Prize in 1996 for his dedication to the creation of a 5-million-acre biosphere reserve in the extremely biodiverse Sierra Madre region of northern Mexico. The Sierra Madre's topography is rugged, carved with gorges deeper than Arizona's Grand Canyon. It is home to endangered species like jaguars, Mexican gray wolves, and thick-billed parrots. Vegetation includes more than 200 species of oak and more species of pine than anywhere else in the world. Native Tarahumara people have developed a complex subsistence agricultural system, which includes cultivation of 15 strains of beans and 20 of corn that are unique in the world. They have honed a refined knowledge of their forest habitat, depending on some 400 wild plants for food and medicine. With its obvious value to science and the national economy (through tourism), this national treasure would, one would think, be conserved automatically. But because of its proximity to the largest drug-using nation in the world, the Sierra Madre is under siege by drug growers and traffickers who are attempting to transform the region into marijuana and opium fields and force the local indigenous Tarahumara people to tend them. Logging operations, some funded by the World Bank, also threaten the forests of the Tarahumara people. Sierra Madre native Bustillos and the grassroots human rights and environmental organization he has founded, CASMAC (Advisory Council of the Sierra Madre), have succeeded in stopping some of the lumber operations and have established one indigenous community reserve, which will become the core of the biosphere project. Fifteen other communities nearby wish to establish their own reserves as part of the biosphere. Bustillos continues to fight for the preservation of his beloved Sierra Madre despite severe injuries sustained in three attempts on his life.

Rachel Louise Carson (1907–1964)

Rachel Carson is the founding mother of today's environmental movement. She studied at Pennsylvania College (A.B. 1929) and

Johns Hopkins University (1932), then began work as an aquatic biologist for the Bureau of Fisheries in 1936. Once that bureau merged with the Biological Survey to become the Department of Fish and Wildlife in 1940, she became its editor of publications. Carson published 12 pamphlets with a strong conservationist message called the Conservation in Action series. She always supplemented this work with independent natural history writing, which was published in magazines and as books. Her book *The Sea Around Us* (1951) was a best-seller and won the National Book Award. Her most famous work, *Silent Spring* (1962), was a poetic yet horrifying wake-up call to a society becoming increasingly dependent upon deadly pesticides. The book's public impact caused the introduction in state legislatures of 42 bills to curb widespread use of insecticides. Carson herself campaigned for many of the bills, and she was awarded the Conservationist of the Year award from the National Wildlife Federation in 1963. But breast cancer (ironically linked to pesticide exposure) claimed her in 1964, and she died before much of the legislation became law. For more information about Carson, write to the Rachel Carson Council, Inc., listed in the Organizations chapter of this book.

Paul Ehrlich (b. 1932)

Trained as an entomologist specializing in butterflies, Paul Ehrlich has been an articulate advocate for environmental conservation for four decades. Shortly after receiving his Ph.D. in entomology from the University of Kansas in 1957, Ehrlich became concerned with global overpopulation. He identified overpopulation as a cause of human misery and of humanity's unsustainable consumption of the earth's natural resources. Ehrlich helped found Zero Population Growth (see Organizations) and wrote *The Population Bomb* in 1968. Ehrlich's concern with this issue continues, but his analysis now views overconsumption by wealthy nations as causing even more harm to the environment than subsistence consumption by poor countries.

Working with his wife, biology researcher Anne Ehrlich, Paul Ehrlich has produced hundreds of articles and more than 30 books about issues related to entomology and others of a more general scope, such as extinction, biodiversity loss, and nuclear war. One of the books the Ehrlichs have cowritten, *Extinction* (see Print Resources), includes the oft-cited metaphor of biodiversity loss as the rivets of an airplane that one by one are being popped off. At a certain point, the airplane will fall apart. Allowing the continued extinction of species, he argues through this analogy, is a gamble we cannot afford.

Ehrlich holds the Bing Professor of Population Studies at Stanford University, where he has worked since 1959. The Sierra Club recognized Ehrlich with the John Muir Award in 1980; he received the World Wildlife Fund's 1987 Medal, and with E. O. Wilson (see below) shared the Crafoord Prize from the Royal Swedish Academy of Science in 1990.

Thomas Eisner (b. 1929)

The parents of world-renowned insect expert Thomas Eisner believe their son learned to walk so he could chase insects in their backyard, and his lifelong fascination with bugs has driven his career ever since. With his wife Maria and Cornell University colleague Jerrold Meinwald, Eisner has cofounded several new entomological fields, including insect physiology, chemical ecology, comparative behavior, biocommunications, and defensive secretions.

Originally from Berlin, Eisner's family moved to Barcelona when Hitler ascended to power in 1933. Then the Spanish civil war erupted, and after a brief stay in France and Argentina, his family emigrated to Uruguay. At 17, shortly after his family moved to New York, Eisner applied to Cornell University but was not admitted, mostly due to his poor English. He went to Champlain College in Plattsburgh, New York, and after two years there, transferred to Harvard, where he earned a Ph.D. in 1955. Eisner was offered a position at Cornell University in 1957 and has been there ever since.

Eisner feels a strong responsibility as a scientist to contribute to conservation efforts. As one of his contributions, he has developed a field he calls "chemical prospecting." It involves searching natural habitats for compounds of use to humanity, such as medicines, pesticides, and chemicals for industrial use. Eisner sees four benefits from chemical prospecting: new jobs for local people, new opportunities for investment, more funds for conservation, and a stronger scientific infrastructure in biodiverse countries. Chemical prospecting has been propounded by environmentalists and rain forest activists and was popularized by such movies as *Medicine Man*.

Eisner is currently the Jacob Gould Schurman Professor of Biology, codirector of the Cornell Institute for Research in Chemical Ecology, and a Senior Fellow at the Cornell Center for the Environment. He is reportedly one of Cornell's most popular lecturers. His infectious enthusiasm for bugs is recorded in the video *The Bug Man of Ithaca* (see Nonprint Resources). Eisner was the 1994 recipient of the National Medal of Science for his "seminal contributions in the fields of insect behavior and chemical ecology and for his international efforts on biodiversity."

Rodrigo Gámez (b. 1936)

The tiny Central American country of Costa Rica is known as a biological treasure chest as well as a vanguard conservationist state. Dr. Rodrigo Gámez has been at the forefront of Costa Rica's conservation movement since the mid-1980s, when he was named presidential advisor on natural resources.

Gámez approached his advisory duties from what he calls the "three pillars of balanced development": environmental concerns, economic needs, and equity issues. This three-pronged approach led him to invent a new organizational blueprint for Costa Rica's already vast national park system. He decentralized the system and reorganized the parks geographically into "conservation areas," each area with its own headquarters. Local residents were integrated into decision making about the parks and were given opportunities to take on active roles in their protection as employees, technicians, or ecotourism entrepreneurs. During this period, Gámez helped found the Neotropical Foundation, which promotes environmentally friendly economic development in the communities closest to national parks.

Gámez founded the Costa Rican Institute of Biodiversity (INBio; see Organizations) in 1989, with $5 million from a debt-for-nature swap (see Glossary and Thomas Lovejoy's biography), and continues with INBio as its director.

Gámez studied agronomy at the University of Costa Rica. He received his master of science degree in plant pathology at the University of Florida in 1961 and his doctorate in plant virology at the University of Illinois in 1967. He worked for three decades in these fields before turning to biodiversity conservation. Gámez has received many national awards, worked with international conservation efforts, and in 1995 accepted the Prince of Asturias Prize on behalf of INBio.

Stephen Jay Gould (b. 1941)

A gift for describing science in terms that laypeople can appreciate has made professor and researcher Stephen Jay Gould a popular guide to his specialties: evolutionary biology, the history of science, and paleontology. Gould's essays eloquently entertain and inform his readers about natural history, warning them of the consequences of human intervention. His essays, published monthly in the magazine *Natural History*, are gathered into books including *Ever Since Darwin* (1977), *The Panda's Thumb* (1980), *Hens' Teeth and Horses' Toes* (1983), *Wonderful Life* (1990), *Eight Little Piggies* (1993; see Print Resources), and several more.

Gould received his undergraduate training at Antioch in Ohio and earned his Ph.D. at Columbia. He has been a professor of both geology and zoology at Harvard University for more than 30 years. One of his major studies was on the land snails of Bermuda and the Bahamas. When exotic snail predators were unknowingly introduced to those islands, the native snails became almost totally extinct. This experience converted him into an ardent conservationist. Gould is also known for many advances in evolutionary theory. One of his most important contributions was his 1972 theory of punctuated equilibrium, which posits that evolution is not a steady process but instead can come in sudden bursts, with rapid change over short periods of time (100,000 years).

Daniel Janzen (b. 1939)

Champion of Costa Rica's megapark system, ecologist Daniel Janzen has been instrumental in designing a new model for national parks in the tropics that includes the varied ecosystems necessary for migratory species and incorporates rural neighbors into the conservation strategy. His commitment to carrying out his vision has made him an effective fund-raiser, convincing U.S. donors to donate millions of dollars to Costa Rican national parks.

Janzen's "baby" is Guanacaste National Park (GNP) in northern Costa Rica, which features the dry tropical forest that has been his subject of study for nearly 40 years. Janzen first began working at Santa Rosa National Park (the park that preceded GNP and sits adjacent to it), where it became clear to him that it would be impossible to carry out his field studies on unprotected land. He often found that between his visits to unprotected study sites, farmers and ranchers chopped down the natural habitat. Though many other tropical biologists escaped this problem by working in privately held reserves, Janzen felt that the conservation of wilderness could not be assured unless the government took responsibility as guardian. Janzen credits this belief to growing up with a father who was for many years the director of the U.S. Fish and Wildlife Service.

During Costa Rica's dry season, Janzen lives in Santa Rosa National Park in a small cement-block house filled with botanical specimens. He returns during the rainy season to his post as biology professor at the University of Pennsylvania, where he has worked since 1976. Janzen studied entomology at the University of California at Berkeley, receiving his Ph.D. in 1965. He first traveled to Costa Rica in the mid-1960s and established a renowned eight-week graduate-level field course in tropical biology for the Organization of Tropical Studies. Janzen received a Crafoord Prize from the Royal Swedish Academy of Science in 1984 for his studies of coevolution and donated half of the $100,000 award to Guanacaste Conservation Area's endowment. Janzen edited *Costa Rican Natural History* (1983), a comprehensive encyclopedia of that country's wildlife, with entries by 174 contributors.

Stephen Kellert (b. 1944)

Professor Stephen Kellert is known for his ground-breaking studies on societal attitudes toward nature. Since the early 1970s, he has surveyed Americans on a variety of topics: public attitudes toward animals, psychological characteristics that differentiate hunters and nonhunters, urbanites' perception of wilderness, attitudes of private owners of forested land toward wildlife management, public attitudes toward energy development in the West, childhood cruelty toward animals by criminals and noncriminals, and insect phobias. The results of studies like these are important to conservationists as they try to convince the American public to support their causes. Kellert's work is also useful for educators seeking to teach younger generations about the importance of biodiversity—even its less appealing forms, such as insects.

Kellert was educated at Cornell and Yale universities and teaches at Yale's School of Forestry and Environmental Studies. His published books include *The Value of Life: Biological Diversity and Human Society* (Island Press, 1996) and *The Biophilia Hypothesis* (with E. O. Wilson, Island Press, 1993). He has received the Distinguished Individual Achievement award from the Society for Conservation Biology and was recognized as Conservationist of the Year by the National Wildlife Federation in 1983.

Winona LaDuke (b. 1959)

Native American activist Winona LaDuke is an articulate spokesperson for the rights of indigenous people to control their own homelands—and the biodiversity they harbor—worldwide. Native peoples have lived sustainably for thousands of years, and LaDuke challenges mainstream white environmentalists who distrust native peoples to manage their lands according to conservationist principles. Residing on the White Earth Reservation in Minnesota, on the ancestral lands of the Ojibwa people, LaDuke manages the White Earth Land Recovery Project, which seeks to buy ancestral lands from their current white owners, piece by piece.

The Harvard-educated LaDuke attributes her activism to her people's spirituality, which views women as the manifestation of Mother Earth and charged with the responsibility of maintaining Mother Earth's health. Currently, the task of protecting the earth requires a defensive posture, since white civilization, acting as predator, sees natural ecosystems as prey. This metaphor can be stretched: indigenous societies and their lands are prey to government and industry, women are prey to men. One-third of native lands in North America are forested; native peoples depend upon them for medicines, baskets, hunting, maple syrup, housing, and fishing. Yet the lumber companies are plundering them for telephone directories and toilet paper.

LaDuke cochairs the Indigenous Women's Network, is on the board of directors of Greenpeace, and publishes *Indigenous Women* magazine. She ran for vice president in 1996 on the Green Party ticket and was recognized by *Time* magazine in 1995 as one of the "Fifty Leaders for the Future."

Aldo Leopold (1886–1948)

Known as the father of environmental ethics, Aldo Leopold's 1949 *Sand County Almanac and Sketches Here and There* advocates a change in "the role of *Homo sapiens* from conqueror of the land-community to plain member and citizen of it." This view was integral to most indigenous societies but new for most European-Americans. It has served as inspiration for many conservation organizations, most directly the Wilderness Society, which Leopold helped found (see Organizations), and has also inspired national legislation such as the Endangered Species Act of 1973.

An Iowa native, Leopold studied forestry at Yale University in the first decade of this century. After graduating, he traveled to New Mexico where he worked his way up to supervisor of Carson National Forest. Leopold wrote a primer called *Game Management* in 1933, which has guided biologists in managing wildlife ever since. In that book, as well as in many articles written for foresters, sportsmen, conservationists, economists, and the general public, Leopold insisted that a respect for all life-forms occurring in a given environment is fundamental for conservation work. This principle challenged the prevailing basis for conservation of that time: the environment should be conserved to assure future generations a ready supply of natural resources.

After many years in New Mexico, Leopold and his family moved back to his home territory and bought a farm in Sand County, Wisconsin. Leopold died while fighting a fire on a neighbor's farm, and *A Sand County Almanac* was published posthumously. Leopold's five children all became scientists.

Thomas Lovejoy (b. 1941)

Thomas Lovejoy has studied the ecology of the Brazilian Amazon since 1965. For decades he has been engaged in the Biological Dynamics of Forest Fragments Project, a giant experiment to determine the minimum size for a viable national park or biological reserve. Consisting of a number of remaining forest patches scattered through a deforested section of the Amazon Basin, this experiment builds on the theory of island biogeography proposed by Robert MacArthur and E. O. Wilson in 1967, which held that the number of species an island can support is determined by the size of the island and its distance from other islands or a mainland. Lovejoy's experiment will continue into the next century, but early observations have shown that diversity definitely declines in small forest islands. The loss of certain species from a patch of forest causes others to disappear as well. For example, the loss of peccaries results in a secondary loss of three species of frogs adapted to breed in the small pools that fill peccary wallows.

In addition to his work in this area, Lovejoy is known for having brought the issue of tropical forest deforestation to public attention and having invented debt-for-nature swaps, whereby a conservation organization buys a portion of a country's foreign debt and gives it back to the country in return for the country's conservation of ecologically valuable land. Currently, Lovejoy is counselor for biodiversity and environmental affairs at the Smithsonian Institution.

James Lovelock (b. 1919)

Dr. James Lovelock is a British scientist who has made important contributions to a variety of fields during his eclectic career, including biology, chemistry, medicine, and physics. Lovelock is most famous as the proponent of the Gaia Hypothesis, which postulates that through increasing or decreasing its absorption and production of atmospheric gases, the biosphere actually regulates the chemical composition of the atmosphere and therefore assures the presence of the right conditions for life to continue.

The Gaia Hypothesis first occurred to Lovelock in the 1960s when he turned his attention to the atmospheres of other planets. Lovelock was sought out by NASA because he had invented a device called the electron capture detector, which could be used to measure chemical levels in the environment. NASA used the electron capture detector to study the atmosphere of Mars when it sent a spacecraft to visit Mars in 1965. In looking for extraterrestrial life on Mars, Lovelock began to study earth as an extraterrestrial observer might and suggested that our atmosphere would be seen extraterrestrially as a constantly self-regulating shield.

Lovelock first published an article describing the Gaia Hypothesis in 1968. The hypothesis melded atmospheric chemistry with ecology and physics. One of the article's editors, microbiologist Lynn Margulis, herself a brilliant, eclectic scientist who regularly strayed from the bounds of tradition, later became an important collaborator in refining the theory. Despite criticism from the scientific establishment, which disapproves of such multidisciplinary tendencies, Lovelock continued writing books and articles detailing Gaia, including *Gaia, A New Look at Life on Earth* (1979), *The Ages of Gaia* (1988), and most recently, *Healing Gaia* (1991). Lovelock counters scientists' criticism by admonishing them that "the ethic of science is based on a fundamental open-mindedness" (*Healing Gaia* 7).

Royalties from his 6 inventions allow Lovelock to work independently out of his home laboratory in the small town of St. Giles on the Heath, England. Lovelock has been honored as Fellow of the Royal Society, Commander of the British Empire, and Associate of the Royal Institute of Chemistry.

Wangari Maathai (b. 1949)

In 1977, Kenyan Wangari Maathai became concerned about nutritional deficiencies among her country's rural people, and after some research, traced the problem to the depletion of fuelwood resources. Because women had to walk so far to fetch wood for their hearth, they had less time to prepare food for their families. So Maathai set up a native species fuelwood tree nursery in her backyard and paid farmers to transplant them to their subsistence farms. This project became Kenya's world-famous Green Belt Movement. In the 20 years since its humble beginnings, 1,500 Green Belt Movement tree nurseries have been established throughout the country, and 10 million trees have been planted. Children are not the only beneficiaries. The movement has also resulted in soil conservation, habitat restoration, and paid work for 80,000 mostly female tree planters. The movement has spread to a dozen African nations.

Born to a family of subsistence farmers, Maathai would probably have become one herself if an older brother hadn't urged their parents to let Maathai study at a special girls' school. Maathai's teachers recognized her brilliance and saw to it that she obtained a scholarship to study in the United States. After receiving a B.A. from Mount St. Scholastica in 1964 and an M.S. from the University of Pittsburgh in 1965, Maathai returned to Kenya, where she earned a Ph.D. at the University of Nairobi. She was the first woman in her country with advanced academic degrees.

Her effective work on behalf of the poor and the environment, and her willingness to speak out against injustice, has threatened the authorities. She has been vilified by the government and imprisoned several times. Internationally, however, her work has been lauded. Maathai received the Woman of the Year award in both 1983 and 1989, the Windstar Award for the Environment in 1989, the Goldman Environmental Prize in 1991, and the Africa Prize for Leadership in 1991.

Russell Mittermeier (b. 1949)

Russell Mittermeier is an internationally recognized expert on primates and reptiles and an influential advocate of biodiversity preservation in the most biodiverse countries of the world. His extensive travels in poor tropical countries have convinced him that poverty is the greatest culprit in environmental destruction. Committed to the preservation of biodiversity, Mittermeier has taken on positions that have allowed him to advise governments worldwide on how to solve their environmental problems. He has worked with the World Conservation Union (IUCN), the World Health Organization, the World Wildlife Fund (WWF), and the World Bank's Task Force of Biodiversity. Since 1989, Mittermeier has been president of Conservation International, a nongovernmental organization that facilitates conservation around the globe. Conservation International (CI) signed the first debt-for-nature agreements in 1987 and currently focuses its conservation efforts in Global Biodiversity Hotspots, the 17 areas of the world that make up 2 percent of the earth's area but harbor 50 percent of its species. Mittermeier is a proponent of a "megadiversity country" protection plan, whereby certain highly biodiverse countries would be declared world protection zones.

Mittermeier studied at Dartmouth and Harvard universities and currently is an adjunct professor at the State University of New York at Stonybrook.

Gary Nabhan (b. 1952)

Desert ecologist and ethnobotanist Gary Nabhan has been described as "one of the most original minds at work in North America" (Terry Tempest Williams), "a seer and celebrant of 'the remaining riches of the living world'" (Gretel Ehrlich), "a naturalist in the full sense of the word, because he has not forgotten the people" (Barry Lopez), and "a biological Studs Terkel" (Wes Jackson). His publications list (including more than 200 articles and 11 books) records his many interests and areas of expertise. His books include *Enduring Seeds* (1989; see Print Resources), an elegy for the disappeared crops and methods of southwest indigenous agriculture; *The Geography of Childhood* (1994), on children's need for wilderness; *Songbirds, Truffles and Wolves: An American Naturalist in Italy* (1993), a travelogue describing the culture and botany encountered on a vacation in Italy; and most recently, *Forgotten Pollinators* (1996; see Print Resources), which describes a secondary effect of the current extinction crisis: the loss of the relationships between flowering plants and their pollinators. Nabhan's academic articles address topics such as use of desert ecosystems as agricultural models, foods native to desert agriculture that help Pima Indians control diabetes, and how gardeners can help reverse degradation of the desert.

Nabhan was a high school dropout but received his B.A. at Prescott College and his M.S. and Ph.D. at the University of Arizona. He cofounded Native Seeds/SEARCH (see Organizations), which seeks to restore the diversity of Native American agriculture and agricultural methods. Currently, Nabhan is science director at Sonora-Arizona Desert Museum in Tucson.

Amooti Ndyakira (b. 1956)

Ugandan journalist Amooti Ndyakira has helped his country steer a course out of the intense environmental ruin of Idi Amin's 1970s dictatorship back to where it can once again live up to its reputation as "the Pearl of Africa." Ndyakira is the only journalist in his wildlife- and natural resource–rich country to address environmental issues. He uses his position at the widely respected independent newspaper, *The New Vision*, to raise public awareness of the fact that natural resources are finite and must be conserved and used sustainably. "Only when people are informed will they be aware, only when they are aware will they take action, and only when they take action will species and the environment be saved," he has said. Examples of his effective feature writing and investigative journalism include a story on the mountain gorillas of the upland forests of Bwindi. During his research there, he discovered that their forest reserve was being plundered by illegal mining, tree cutting, and poaching. The public concern resulting from his article led the Ugandan Parliament to upgrade the reserve's status to national park, providing stricter protection to the gorillas and their habitat. Ndyakira also discovered that although the government had agreed to join the Convention on International Trade in Endangered Species (CITES), the presiding minister would not sign necessary documents. Through concerted pressure on the minister, Ndyakira finally succeeded, in 1992, in convincing the minister to sign. But even after the country had signed CITES, Ndyakira discovered much illegal wildlife trade still taking place. Putting himself in great danger, he assisted in a sting that exposed the smuggling of chimpanzees and African great gray parrots. Much to the shock of the Ugandan people, participants in this and other illegal smuggling operations included airport personnel, game officers, businessmen, foreign diplomats, and foreign relief workers. This heightened awareness has made the authorities work harder to prevent further smuggling.

Ndyakira received a Goldman Environmental Prize in 1996 for his bravery.

Steve Packard (b. 1943)

Through his work with the North Branch Prairie Project outside of Chicago, Steve Packard has inspired a nationwide grassroots ecological restoration movement. He is not an academically trained scientist. But far from being a detriment, his varied experience has been an advantage for his work. His eclectic background as a child natural history buff, member of a film collective that made political documentaries, and anti–Vietnam War movement organizer has allowed him a flexibility and creativity not often found in professionals whose interests are restricted to their discipline.

In 1975, with the war over and no clear idea about what to do next, Packard returned to his childhood passion for nature and began exploring the wildlands surrounding Chicago. To his dismay, many of the "nature preserves" protected by the Illinois Forest Preserve District had become trash- and weed-filled wastelands. Packard had read a description by prairie ecologist Robert Betz of these ecosystems in their full glory, and he wondered what it would take for this ecosystem to regain its beautiful original state. One day, he experienced a vision of each prairie with its own "congregation" of restorationists. Drawing on his organizing experience, he quickly gathered these volunteers and gained permission from the Forest Preserve District to work on the sites. Little by little, with profound ecological discoveries, lots of hard work, plenty of controversy, and difficult setbacks, the savanna and prairie ecosystems in the area are looking more like they used to. Over 20 years after he began this project, more than 5,000 volunteers work on 202 sites throughout Illinois, on a total of 27,000 acres. Packard currently works with the Nature Conservancy's Illinois office and speaks throughout the country. His work, and some of the many similar projects in North and Central America, are described in William K. Stevens's *Miracle under the Oaks*, reviewed in the Print Resources chapter of this book.

Paula Palmer (b. 1948)

As a wandering backpacker in 1973, Paula Palmer arrived in Costa Rica's Talamanca region with no intention to stay. It rained during her entire stay in the village of Cahuita, and she spent her time helping the local English-speaking Afro-Caribbean ladies who cooked for widowers, bachelors, and visitors like herself. Soon one of the hostesses asked Palmer to tutor her grandchildren in English. When Palmer reappeared with a notebook and pen, there were 37 children gathered in the backyard. Within a week the local community elected a school board and hired Palmer as director of a new English school. Since there were no appropriate teaching materials available, Palmer began recording stories from Talamanca's oldest Afro-Caribbean residents. She would bring the transcriptions to class, and her students learned English by reading stories from their community's elders. Parents too read the tales, and eventually the school board asked Palmer to gather the stories into a book. *What Happen*, which came out in 1978, documents the self-reliant and sustainable lifestyle led by Afro-

Caribbean people who live in this remote area of the country. Palmer was later invited by the KéköLdi indigenous communities south of Cahuita to help them write their own book as well. The result is *Taking Care of Sibö's Gifts*, reviewed in the Print Resources chapter of this book.

Palmer's work follows the guidelines of "Participatory Action Research," as established by the Participatory Research Network of the International Council for Adult Education, which entails helping a community undertake research on its own culture, history, politics, and ecology, and then urging community members to use what they have learned when making decisions about their future. In the case of *What Happen*, the Afro-Caribbean residents of Talamanca were given a new sense of the uniqueness of their cultural history. *Taking Care of Sibö's Gifts* inspired the KéköLdi people to initiate a captive-breeding project for the endangered green iguana, open a cultural center where elders teach children the indigenous language and traditional knowledge, and share their knowledge of the rain forest with outsiders through guided tours. For their efforts, the KéköLdi people were given the Richard Evans Schultes prize for ethnobotany and biodiversity conservation.

Palmer, who has a B.A. in English literature from the University of Colorado and an M.A. in sociology from Michigan State, moved back to her hometown of Boulder, Colorado, after 20 years in Costa Rica. She currently serves as the executive director for Global Response (see Organizations) and as editor for the health and environment section of *Winds of Change*, an American Indian quarterly.

David M. Raup (b. 1933)

David Raup has performed important research on ancient evolution and extinction that provides clues to the current puzzle of how biodiversity has developed and what threatens it. His name and descriptions of his work appear frequently on the science pages of the *New York Times*. One of his most recent projects was a decade-long measurement study of fossil mollusks that culminated in the refutation of what was long held to be a truism in evolution: organisms get bigger through time.

In the introduction to one of Raup's five books (*Extinction, Bad Genes or Bad Luck?*; see Print Resources), Stephen Jay Gould characterizes him as a "brilliant colleague" and "the master of quantitative approaches to the fossil record"—one whose open mind allows him to give serious consideration to ideas that more rigid scientists shut out. His openness allowed him to investigate the Alvarez theory that meteorites hitting earth caused extinctions and write persuasive articles resulting in widespread acceptance for the theory.

Raup did his undergraduate work at the University of Chicago and his graduate work at Harvard University, where he received his Ph.D. in 1957. He is currently professor of invertebrate paleontology at the University of Chicago.

Richard Evans Schultes (b. 1915)

One of the most distinguished ethnobotanists in the world and a pioneer in his field, Richard Evans Schultes is an internationally recognized expert in the botany of rubber trees, medicinal plants, and hallucinogens. He worked in the rain forest for more than 40 years and spent 14 years living with Amazon tribes. He has collected more than 24,000 plants of potential medicinal and food use. From his experience with human inhabitants of the rain forest, Schultes became convinced that chemical prospecting for potential medicines is most effective if guided by the knowledge of indigenous people. Indigenous shamans using healing practices that have been passed down from generation to generation have developed a deep understanding of medicinal substances in their habitats. Schultes warns about the fragility of this knowledge base: "When our civilization arrives with roads, missionary activities, commercial interests, tourism, or otherwise, the products of our culture are rapidly adopted and, even in one generation, replace what has for hundreds of years been part of their culture." An organization named for one of his books, the Healing Forest Conservancy, grants an annual Richard Evans Schultes Award to a scientist, practitioner, or organization that has made an outstanding contribution to ethnobotany or indigenous peoples' work related to that field.

Schultes has published more than 400 technical papers and nine books, including *The Healing Forest* (with Robert Raffauf, 1990) and *Vine of the Soul* (1992). Schultes is a Professor Emeritus at Harvard University. He has received a Gold Medal from the World Wildlife Fund, the Tyler Prize for Environmental Achievement, and the Linnean Gold Medal.

Vandana Shiva (b. 1952)

In 1982, in the midst of a successful career as researcher at the Indian Institute of Management, Vandana Shiva abandoned particle physics and turned her focus to the worldwide degradation of biodiversity and the rapid expansion of biotechnology. Although the two issues may not appear to be related, Shiva draws clear connections between them. Shiva insists that biodiversity occurs not only in undeveloped natural areas but also in the intricate production pattern of a traditional agricultural village. A wide variety of crops is grown, some for direct human consumption, others as fodder for livestock; waste products like cow manure and unused grain stalks replenish the soil with lost nutrients. Biotechnology threatens biodiversity when it disrupts this style of agriculture. "Desirable" crop characteristics, such as high productivity, are bred by seed companies into hybrids, and the seeds are widely distributed in traditional agricultural countries such as India. Although the commercial seeds may produce more, plant for plant, than the traditional seeds, traditional varieties may be preferable in other ways. Farmers may care about the quality of a plant stalk for fodder. Or they may have bred drought or pest resistance into a crop over generations. Farmers using commercially bred seeds lose the side benefits of their traditional varieties and agricultural methods. Even more serious is that when seed companies discover varieties developed over generations by traditional farmers, they can take out an international patent on their genes, effectively robbing the farmers of the legal use of their heirloom seeds.

Shiva has produced more than a dozen books, including *Biopiracy* (1997) and *Biodiversity: Social and Ecological Perspectives* (1991; see Print Resources). She directs the Research Foundation for Science, Technology and Natural Resource Policy, a network of researchers who specialize in sustainable agriculture and development, and is the ecological advisor to the Third World Network, which gives voice to people of poor countries. She has been honored with a UN Global 500 Award and in 1993 was given Sweden's Right Livelihood Award, known as an alternative Nobel Prize.

Michael Soulé (b. 1936)

This unconventional, interdisciplinarian mind is responsible for founding a new field, conservation biology, which links biology to other fields—such as economics, international development, and environmental policy—that play a role in conserving what biologists study.

Soulé began his career by studying the thermoregulation and biogeography of lizards, which led him to evolutionary theory and ecological patterns. Shortly after obtaining the dream job for most academics, a professorship at the University of California–San Diego, Soulé quit, gave away his hand-built house, and joined a Buddhist center, where he served as director for six years. His love for biology never ceased, however. Soulé returned to academia to found and serve as first president of the Society for Conservation Biology. He took over and revitalized the ailing environmental studies program at the University of California–Santa Cruz, which had been the first of its kind. Soulé currently works independently out of his home on the western slope of Colorado.

Edward O. Wilson (b. 1929)

Harvard University entomologist Edward O. Wilson can probably be credited for making biodiversity into a household word. During his long and productive career, he has pioneered new fields of biology and has written for both specialists and the general public. His eloquence has made him a popular spokesperson for biodiversity.

Wilson's lifework is evidence for his theory that the best scientists are driven by an artist's passion for their work. "You start by loving a subject—the odds are that the obsession will have begun in childhood. The subject will be your lodestar and give sanctuary in the shifting mental universe," he writes. Growing up as a lonely, solitary boy in the Deep South of the 1930s, Wilson immersed himself in the fantastic ecology of the Gulf Coast. When a fishing accident injured one of his eyes, permanently damaging his vision, Wilson specialized in a life-form requiring a smaller visual field: insects, specifically ants. Wilson performed the first inventories of ants through most of the southern Pacific, and during these explorations of tropical islands, he began to suspect that the chaotic display of life in the tropics concealed a very orderly distribution of species.

After sharing this inkling with ecologist Robert MacArthur in 1962, Wilson and MacArthur together developed the theory of "island biogeography," which related the number of species on an island—either an oceanic one or an area surrounded by unlike ecosystems—to its area and its distance from the mainland (or in the case of a nonoceanic island, a similar ecosystem). This theory was important then and has gained further significance recently as deforestation and habitat alteration create shrinking islands of natural habitat in seas of deforested pastures or suburban residential expansion.

Wilson has published hundreds of articles and several books for specialists, including *The Theory of Island Biography* (1967, with Robert MacArthur), *The Insect Societies* (1971), *A Primer of Population Biology* (1973, with W. H. Bossert), *Sociobiology* (1975), *Caste and Ecology in Social Insects* (1978, with George Oster), and *The Ants* (1990, with Bert Holldobler). He has also written such classics for lay readers as *On Human Nature* (1978), *Biophilia* (1984), *The Diversity of Life* (1992), and *The Naturalist* (1994). Several of his books are reviewed in the Print Resources chapter of this guide.

Wilson has been awarded a Pulitzer Prize (for *On Human Nature*). He received the 1977 National Medal of Science, and in 1990 he shared with Paul Ehrlich (see above) the Royal Swedish Academy of Science's Crafoord Prize. He was awarded the WWF Gold Medal in 1990, and in 1993 the Japanese gave him the International Prize for Biology.

Statistics, Illustrations, and Documents

The material included in this chapter is meant to give an overview of some of the main facets of biodiversity, to highlight the research that has elucidated its mysteries, to outline the most pressing threats, and to sketch out proposed solutions to them. But to researchers who hunger for more: do not limit yourselves to what this brief chapter contains. There are literally thousands of useful charts, tables, maps, and diagrams in the atlases and reports listed in the Print Resources chapter.

Facts and Statistics

The following tables describe biodiversity quantitatively and qualitatively. They provide such information as how many species have been identified, how useful some are to humans and other earth dwellers, where habitat is most endangered, and what parts of the world are doing the best job conserving it.

The Numbers of Species

How many species may exist is, at this point, a question with many answers. Until 1982, it was commonly thought that there were up to

5 million species on earth. But when entomologist Terry Erwin published his findings that a single tree could host up to 1,500 species of beetle, many of them living only on that species of tree, estimates immediately began to balloon. Erwin believes that up to 30 million species of arthropods may inhabit the tropics alone. Currently, mainstream scientists say that there may be anywhere from 50 to 100 million species.

TABLE 4.1
Numbers of Described Species of Living Organisms

Kingdom and Major Subdivision	Common Name	Number Described
Virus	Viruses	1,000
Monera	Bacteria and blue-green algae	4,760
Fungi	Fungi and molds	46,983
Algae	Algaes, dinoflagellates, and euglenoids	26,900
Plantae	Nonflowering plants	17,750
	Flowering plants	250,000
Protozoa	Protozoans	30,800
Animalia		
Invertebrates:	Sponges	5,000
	Jellyfish, corals, jellies	9,000
	Flatworms	12,200
	Roundworms	12,000
	Earthworms and relatives	12,000
	Mollusks	50,000
	Echinoderms	6100
	Arthropods:	
	Insects	751,000
	Others	123,161
	Other invertebrates	9,300
Total invertebrates		989,761
	Tunicates	1,250
	Acorn worms	23
Vertebrates:	Lampreys and other jawless fishes	63
	Sharks and other cartilaginous fishes	843
	Bony fishes	18,150
	Amphibians	4,184
	Reptiles	6,300
	Birds	9,198
	Mammals	4,170
Total vertebrates		42,908
Total, all organisms		1,412,135

How many species are known to science is an entirely different question. Table 4.1, a condensed version of the tables devised by biologist E. O. Wilson for *Biodiversity* (Washington, DC.: National Academy Press, 1988; (see pp. 49–51) and the compilers of *Conserving the World's Biological Diversity* (Gland, Switzerland, and Washington, DC: IUCN, WRI, CI, WWF, and the World Bank, 1990, p. 18), lists the numbers of species that had been identified up to 1988 and 1990, respectively.

Medicines Derived from Natural Sources

Pharmacologists estimate that one-quarter of all the medicines used in the United States contain active ingredients extracted from plants. Americans spend more than $8 billion on these medicines. Worldwide, almost 75 percent of all people depend on medicines—either pharmaceuticals or traditional medicines—from plant and animal sources. Table 4.2 is a representative listing of medicines derived from natural sources.

Services Provided by Ecosystems

A new argument for conservation that is beginning to supersede the species-by-species approach is that intact ecosystems provide a great number of useful services. Conservationists argue that without these services, life on earth would soon collapse due to a lack of air, water, and soil nutrients. Table 4.3 lists the major services provided by ecosystems.

Protected Areas

Conservationists believe that biologically important areas representing all types of ecosystems should be protected worldwide, not just in countries that can "afford" to set land aside. Globally, protected areas have expanded by 80 percent since 1970, with poor countries responsible for about two-thirds of that amount. International conservation organizations have assisted in the effort, offering economic and technical support for conservation. However, the amount of land protected needs to grow three times larger than it is currently to ensure the long-term viability of the earth's major ecosystems. Table 4.4 shows the percentage of land protected in 14 geographical regions of the world.

TABLE 4.2
Medicines That Originate from Natural Sources

Source	Property
Aloe	Juice soothes skin burns
Ancistrocladus korupensis vine from Cameroon	Protects human cells from HIV virus
Arnica montana flower	Heals abrasions and bruises
Bee venom	Soothes arthritis
Cashew nut oil	Fights tooth decay, inhibits bacterial growth
Chamomile	Mild sedative, digestive
Cone flowers (Echinacea)	Immunological system booster
Dandelion	Soothes heartburn, digestive, helps liver and gallbladder ailments
Dogfish aminosterol	Antiparasitic, antifungal, antiprotozoan, antibacterial
Ergot mold	For pain from migraines, childbirth
Fer-de-lance snake venom	Relieves hypertension
Feverfew	Relieves fever and headache
Foxglove (digitalis)	Heart disease
Fungus (unnamed, from Norway)	Contains cyclosporine, which prevents white blood cells from from attacking a new organ transplant
Garlic	Antiseptic, reduces cholesterol levels and hypertension
Ginkgo	Heart and lung ailments, asthma, Alzheimer's disease
Ginseng	Strengthens immune system
Goldenseal	Antiseptic, antiyeast, antibacterial, antiparasitic
Gotu kola	Strengthens veins and capillaries, aids brain function
Himalayan mayapple	Kills intestinal worms, laxative
Hyssop	Relieves respiratory ailments
Indian snake root (reserpine)	Sedative, cures hypertension and mental illness
Lavender	Sedative, soothes intestinal ills
Licorice root	Soothes ulcers, anti-inflammatory
Mahuang (ephedra)	Decongestant
Madagascar periwinkle	Cures childhood leukemia, Hodgkin's disease
Mayapple	A natural extract called podophyllotoxin cures testicular cancer; also used for herpes, influenza, and the measles
Mexican yam	Used to make steroids
Neem tree	Leprosy, diabetes, ulcers, skin disorders, constipation
Opium poppy	Natural painkiller and mental stimulant
Osha (bitterroot)	Cures headaches and dehydration
Pacific yew tree (taxol)	Treatment of ovarian and breast cancers
Poison arrow frog	Painkiller
Saint John's wort	Anti-inflammatory and antitubercular
Tangerine peel	Promotes digestive function
Trillium	Helps wounds to heal
Turmeric	Relieves allergy symptoms and inflammations
White willow bark (aspirin)	Painkiller
Wild cherry bark	Soothes coughs

Sources: Krasemann, Stephen J., and Noel Grove, *The Nature Conservancy: Preserving Eden* (New York: Harry N. Abrams, 1992); Lovejoy, T. E., "Quantification of Biodiversity," in *Biodiversity: Measurement and Estimation*, D. L. Hawksworth, editor, 81–87 (London: Chapman & Hall, 1995); Facklam, Howard, and Margery Facklam, *Plants: Extinction or Survival* (Hillside, NJ: Enslow Publishers, 1990); Wallace, David Rains, *The Quetzal and the Macaw: The Story of Costa Rica's National Parks* (San Francisco, CA: Sierra Club Books, 1992); "The Healing Power of Plants" (Henderson Museum at the University of Colorado, Boulder; exhibit September 1996–June 1997); Myers, Norman, *Gaia: An Atlas of Planet Management* (New York: Doubleday, 1984).

TABLE 4.3
Goods and Services Provided by Ecosystems

Ecosystems:

- catch, store, and filter water
- regulate temperature, moisture, light, filtration of ultraviolet rays and other radiation
- cycle nutrients such as carbon, hydrogen, sulfur, oxygen, and nitrogen
- build and maintain soil, and protect it from erosion
- buffer human settlements from storms, floods, and other disturbances
- decompose waste products
- control pollution by breaking down contaminants
- maintain habitat for local and migratory wildlife
- provide habitat for pollinators
- provide breeding, nursery, and feeding grounds
- serve as gene banks for continued evolution of organisms
- are a source of biological controls for agricultural pests and diseases
- are a source of raw materials for building, clothing, energy, and fuel
- are a source of food, medicines, biochemicals, minerals, genes, ornaments
- provide the source for human livelihoods: fishing, agriculture, aquaculture, hunting, forestry, tourism, science, conservation
- provide recreation opportunities: "being in nature," hiking, swimming, boating, fishing, insect collecting, photography, sightseeing
- are sacred sites and provide a divine model for many religious and cultural traditions

Sources: Saunier, R. E., *Conservation and Management of Natural Resources for Sustainable Development* (Washington, DC: Organization for American States, 1995); Heywood, V. H., editor, and R. T. Watson, chair, *Global Biodiversity Assessment* (Cambridge, UK: Cambridge University Press, 1995); Worldwatch Institute, *State of the World 1997* (New York: W. W. Norton, 1997).

Table 4.4. Protected Areas around the World

	Percentage of Land Area Protected as:		
Region	Parks/Monuments/Reserves Allowing Recreational, Subsistence, or Scientific Use Only	Areas Where Sustainable Use of Resources Is Allowed	Total
Antarctica	0	0	0
Australia	4.1	8.1	12.2
Caribbean	3.9	5.6	9.5
Central America	6.7	2.3	9.0
East Asia	0.6	5.2	5.8
Europe	2.3	8.6	10.9
North Africa/Middle East	1.4	1.5	2.9
North America	7.6	4.9	12.5
North Eurasia	2.0	1.1	3.1
Pacific	7.9	0.4	8.3
South America	3.8	2.5	6.3
South and Southeast Asia	3.2	2.7	5.9
Sub-Saharan Africa	3.4	2.4	5.8

Source: Heywood, V. H. editor, and R. T. Watson, chair, *Global Biodiversity Assessment* (Cambridge, UK: Cambridge University Press, 1995), 983, 985.

Global Biodiversity Hotspots

Conservation International, an organization that focuses on the most biodiverse and endangered countries of the world, has assembled a list of 17 "Global Hotspots"—together comprising only 2 percent of the earth's land area, but harboring half of the world's species (see Table 4.5).

TABLE 4.5
Global Biodiversity Hotspots as Identified by Conservation International

- The Tropical Andes
  - Venezuela
  - Colombia
  - Ecuador
  - Peru
  - Bolivia
- Madagascar
- The Cape and Western Cape Floristic Region of Southern Africa
- The Antilles
- Western Sunda
  - Indonesia
  - Malaysia
  - Brunei
- Eastern Sunda
  - Indonesia
- The Philippines
- The Atlantic Forests of South America
  - Brazil
  - Paraguay
  - Argentina
- Brazil's Cerrado
- The Darien Gap and the Chocó
- Mesoamerican Forests
- Polynesia and the Micronesian Islands
- Southwestern Australia
- The Eastern Mediterranean Region
- The Western Ghats of India and Sri Lanka
- New Caledonia
- The Guinean Forests of West Africa

Illustrations

Figure 4.1 Defensive Adaptations

Plants and animals have developed a variety of adaptations that help them defend themselves from predators. Figure 4.1 (page 99) depicts three types of defensive adaptations.

Protective physiology refers to features that offer physical protection. The illustration shows thorns that deter herbivores (top and left borders), hard shells that shield the soft bodies of mollusks (lower corners), and adaptations for wind dispersal of seeds that maple trees have evolved. Some trees have evolved seeds encased in a hard shell to protect them from predators; others entice dispersers by offering tasty fruit, which an animal eats before regurgitating or defecating the seed, which is unharmed in the process. But the maple tree, whose winged seeds are pictured at the center of the drawing, above the lower triangles, has a soft seed that is vulnerable to predators. Its wind-catching feature reduces the chances that a predator will destroy its seeds before they begin their journey.

Protective coloration refers to camouflage that diminishes the probability of detection. The eggs of gull-billed terns (upper left corner) and snowy plovers (upper right corner) both rest directly on the ground, so they must blend well with their surroundings of pebbles, sand, and broken shells. Thorn-mimic treehoppers (on the thorny branch between the eggs) are easily overlooked when motionless. Sometimes protective coloration is designed to startle and intimidate. When the sphinx moth is at rest during the day, the camouflage pattern of its upper wings allows it to remain undetected against tree trunks (center). But if disturbed, the Sphinx moth quickly reveals its lower wings, with their glaring eyespots, designed to mimic an owl-like predator (below top thorny border).

Chemical defenses and warning coloration are another deterrent to predators. Many animals secrete chemical toxins to render themselves unpalatable to predators. They advertise their unpalatability with bright colors and graphic patterns; vivid orange, red, yellow, and blue often mean "Yuck!" to predators. Poison-dart frogs (left, above lower triangles) are found throughout Central and South America in a brilliant array of colors and patterns that alert possible predators to their toxicity. Monarch butterflies also depend on chemical defenses. Many plants manufacture harmful compounds to discourage herbivores, but some insects have adapted to take advantage of this strategy. Monarch caterpillars feed on milkweed plants (left triangle), storing the plants' toxic alkaloids in their own bodies. The alkaloids provide chemical protection to both the caterpillars and the resultant butterflies (center triangle), since birds have learned that Monarchs taste bad. The viceroy butterfly (lower right) tries to fool birds into believing that it too is unpalatable by almost perfectly mimicking the monarch (look for the slight differences in the veins and pattern of the lower wing). This type of mimicry—a palatable species adapting to look like an unpalatable species—is called Batesian mimicry. The spiny creature crawling up the right border is a hickory-horned devil, the caterpillar of the regal moth. The body is greenish and the large horns are orange with black tips.

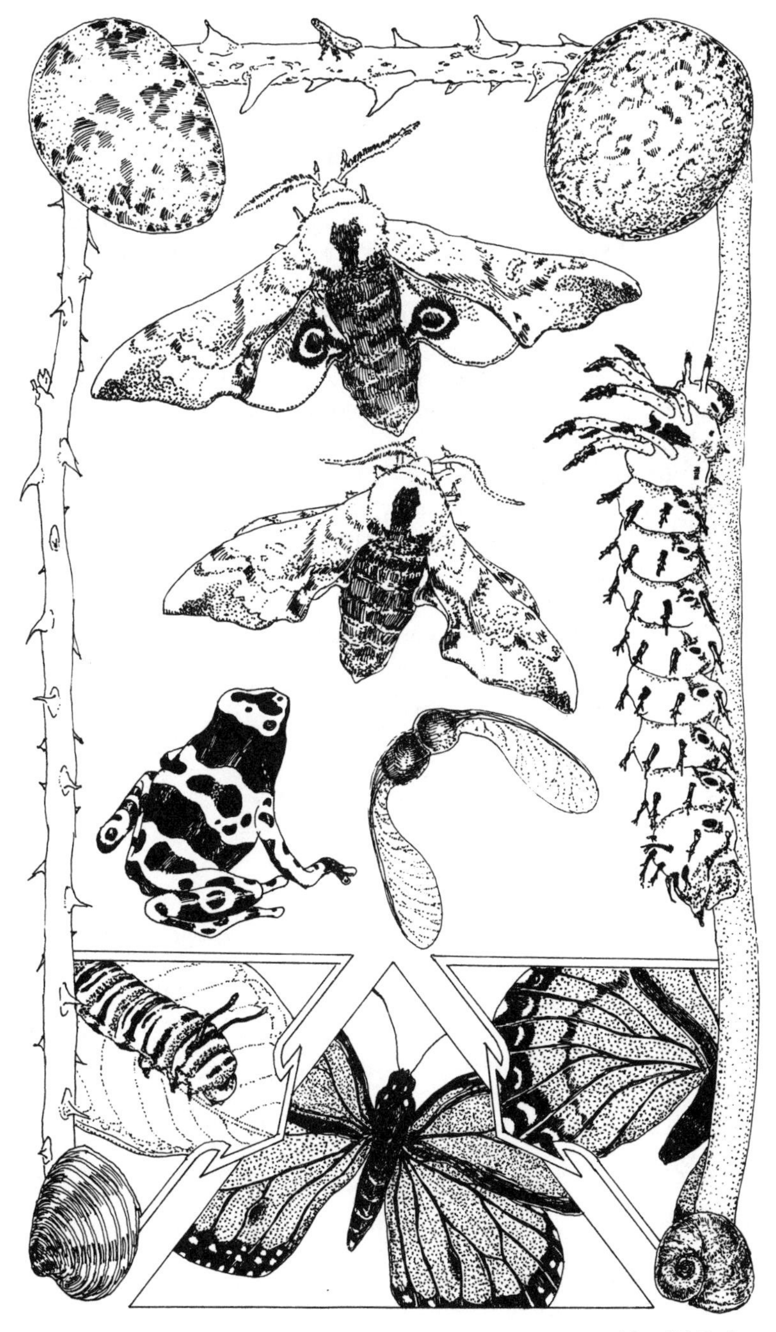

Meredith Broberg

Figure 4.2 Marine Food Web

This drawing illustrates the complex food web of marine ecosystems. Arrows connect predators to their prey. The top box shows photosynthesizing phytoplankton, the primary producers that form the base of the food web. The phytoplankton depicted include green algae, diatoms, and dinoflagellates. The box just below the phytoplankton shows zooplankton, including the larvae of fish, mollusks, and echinoderms—spiny sea–bottom dwellers such as starfish and sea urchins. These graze on phytoplankton. Zooplankton is consumed by baleen whales (upper right), which cruise the upper realms of the ocean, straining plankton through the hornlike plates that fringe their mouths. Most smaller fish are zooplankton feeders, but herrings (the *clupea* species is shown on the upper left) consume phytoplankton as a significant part of their diet. Herrings and other planktonic fish are eaten by seabirds (top center) and larger fish such as cod and haddock (lower right). Seals (mid-left) prey on planktonic and predatory fish. The killer whale eats whatever it wants. Detritus feeders (box on lower left) take advantage of the continual rain of organic matter that sinks from above. Sea cucumbers and brittle stars feed on the rich seafloor mud, extracting bacteria and other organic particles.

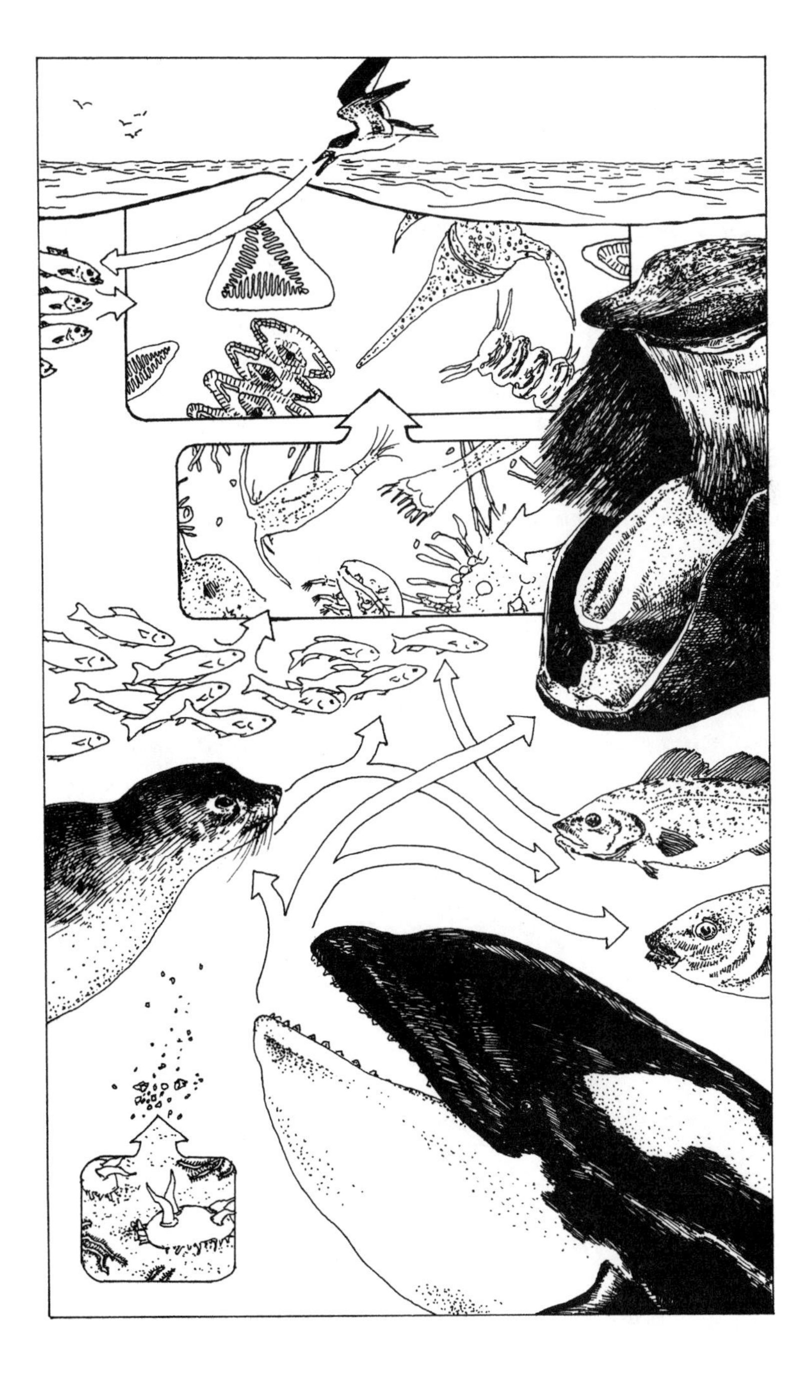

Meredith Broberg

Figure 4.3 Nutrient Cycles

Nutrients such as carbon, oxygen, sulfur, nitrogen, and methane are cycled through ecosystems. The upper left inset illustrates the three modes through which carbon is released into the atmosphere: decomposition, respiration, and burning. It is reabsorbed by plants, which use it in their process of photosynthesis.

The larger diagram details the modes of carbon cycling. Photosynthesis, a process that employs water absorbed by plant roots (lowest arrow), carbon absorbed from the air into the stomata (little openings on leaves), and sunshine, allows the plant to generate sugar to fuel its growth. The by-products of photosynthesis are water and oxygen. Animals such as the caterpillar on the leaf to the left and the aphids on its stem prey upon the plant, converting its carbon into carbohydrates, a type of energy they can use. The animals that prey on the herbivores absorb those carbohydrates, and the carbon cycling web expands with each ensuing predator. When a plant or an animal dies and decomposes (for example, the dead leaf on the lower left), it releases its stored carbon back into the ecosystem.

In addition to the carbon cycled through food consumption and decomposition, carbon is cycled by respiration. Animals breathe in the oxygen produced during photosynthesis and breathe out carbon dioxide, which can be absorbed immediately by plants.

Meredith Broberg

Documents

The documents included in this section either identify crises that must be solved quickly (the Forgotten Pollinators Campaign, the Union of Concerned Scientists' World Warning to Humanity) or propose solutions (Caring for the Earth, National Biodiversity Institute Documents, Convention on Biological Diversity). All of these documents have inspired a global response from experts in the field, but the documents are so loaded with suggestions and ideas that they deserve to be read by a much larger audience.

The Forgotten Pollinators Campaign

During the spring of 1995, the following urgent appeal was issued for recognition and mitigation of a poorly understood problem that is quickly reaching the crisis stage—the disappearance of pollinator-plant relationships caused by widespread habitat degradation. This call was formulated by an advisory board for the Forgotten Pollinators Campaign, spearheaded by the Arizona–Sonora Desert Museum. Sponsors include the Desert Museum, Bat Conservation International, Center for Plant Conservation, National Wildflower Research Center, Sonoran Arthropod Studies Institute, and Xerces Society. (Reprinted with permission of The Forgotten Pollinators Campaign, c/o Arizona-Sonora Desert Museum, 2021 N. Kinney Rd., Tucson, AZ 85743.)

A Call for a National Policy on Pollination

As scientists and educators, food producers and consumers, we are concerned that a basic fact of life—our dependence on the link between plants and their pollinators—is poorly understood by policymakers, the public, land managers, and the food industry. We have joined together to urge further appreciation and appropriate action regarding the following principles:

1. *The Future of Our Farms Depends on Pollination*. The pollination of plants, which is often mediated by animals, is necessary for seed set, fruit yields, and reproduction of most crops. Animal pollinators also serve a diversity of dominant trees and herbs in wildlands. Nevertheless, pollination remains one of the weakest links in our understanding of how ecosystems

function and how crop yields can be assured. Population ecology should be taught in every agricultural program and land grant college in the United States.

2. *We Need to Appreciate the Benefits That a Diversity of Pollinators Provides.* Pollination services are provided by a diversity of animals in addition to the domestic honeybee. And yet the "free" pollination services provided to food and forest crops seldom enter into government statistics on the value of protecting wild species or the "costs" of maintaining agricultural yields. A complete inventory of pollinators of crops and keystone plant species in wildlands should be compiled by the National Biological Service and USDA-ARS [United States Department of Agriculture-Agricultural Research Service] and their counterparts in other countries.
3. *Honeybees Are in Decline.* Since 1990, U.S. beekeepers have lost one-fifth of all their domestic managed honeybee colonies due to the arrival of Africanized honeybees and abandonment of hives, the spread of parasitic mites and diseases, and the withdrawals of honey subsidies that formerly helped support the industry. Wild pollinators must now take up some of this slack in pollinating crops. Collectively, wild and domestic bees provide pollination services 40 to 50 times more valuable than the market price of all honey produced in the United States. The USDA, Mexico's SARH [Secretaria de Agricultura y Recursos Hidráulicas], and Agriculture Canada should invest more in their non-*Apis* bee programs and provide support in order to stabilize the honeybee industry and encourage diversification in pollination industries.
4. *All Pollinators Require Protection from Toxins and Land Degradation.* Maintenance of wild pollinators for crops requires habitat set-asides or greenbelts near agricultural fields. Since wild pollinators are often more vulnerable to pesticides and herbicides than are domestic honeybees, the use of toxic chemicals must be carefully controlled near their nesting and foraging sites. Those who apply pesticides should be better trained in monitoring the health of pollinator populations. When given a choice, they should use insecticides that are known to be less toxic for bees.

5. *Habitat Fragmentation Is a Major Threat to Pollinators.* Although its effects on native pollinators are not completely understood, habitat fragmentation may be reducing their populations due to loss of nesting habitats, indiscriminate use of rangeland and agricultural chemicals, and elimination of host plants and nectar sources. As habitat patches ("islands") become smaller, they may become insufficient to support pollinators through the mix of plants they require. Populations of pollinators should be closely monitored and habitat fragmentation trends mapped in order to determine specific causes of pollinator decline and convince land-use planners of the need to zone greenbelts and corridors to ensure pollinator survival.
6. *Fewer Pollinators Ultimately Means Fewer Plants.* Where habitat fragmentation and pesticides have reduced the populations of pollinators, plants will eventually suffer low reproductive success. The pollination ecology of many plants has barely been studied, even though it may be critical to keeping some of these plants from extinction. We must earmark support for studying these interactions and for including pollinator nesting and foraging areas in critical habitat designations.
7. *Endangered Species Protection Need Not Be Incompatible with Food Protection.* Pesticides and herbicides must not be sprayed in the immediate vicinity of endangered plants, rare pollinators, and their habitats. Nevertheless, the "spraying setback distances" being set by the EPA (Environmental Protection Agency) and USFWS (U.S. Fish and Wildlife Service) have been arbitrarily determined without detailed knowledge of specific plant/pollinator relationships. Such arbitrary determinations not only frustrate ranchers and farmers, but may fail to protect the species they are meant to conserve. Before implementing such setbacks, on-site determinations need to be made by pollination ecologists familiar with the species involved.
8. *Plants and Pollinators Both Need Protected Habitats.* In a few cases, pollinators have declined to the extent that the economically important plants they formerly served are suffering reduced seed set and fruit yields. In other cases, the decline of certain plants has triggered the decline of pollinators that specialize on them. These reciprocities deserve special attention and support to

reverse these trends and restore the relationships. When necessary to keep them from extinction, critical habitat should be set aside for both plants and their pollinators.

9. *Migratory Pollinators May Require International Protection.* Although the reproduction of plants is occasionally reduced by lack of pollinators, this condition has been aggravated by human activity in the landscapes they inhabit and the severing of migratory corridors. Scientists must carefully monitor plant/pollinator changes and assess global trends.
10. *Pollination Is a Threatened Ecological Service.* The loss of biological diversity means more than simply counting the declining number of species. It also implies the extinction of relationships or disruption of ecological processes, such as pollination, upon which we all depend. One in every three mouthfuls of food we eat depends on pollination by bees and other animals to reach our kitchen tables. And yet, thousands of pollinating animal species are globally endangered. Additional habitat conservation, monitoring, research, and ecological restoration will be required to reverse these far-reaching global trends.

Union of Concerned Scientists' World Warning to Humanity

This warning was written and spearheaded by Union of Concerned Scientists (UCS) chair Henry Kendall in November 1992 and was signed by more than 1,700 of the world's leading scientists, including the majority of Nobel laureates in the sciences. The UCS has a history of drawing public attention to serious global threats. The threat posed by a destroyed natural environment may be the most serious yet. (Reprinted with permission of the Union of Concerned Scientists.)

World Scientists' Warning to Humanity

Introduction

Human beings and the natural world are on a collision course. Human activities inflict harsh and often irreversible damage on the environment and on critical resources. If not checked, many of our current practices put at serious risk the future that we wish for human society and the plant and animal kingdoms, and may so alter the living world that it will be unable to sustain life in the manner that we know. Fundamental changes are urgent if we are to avoid the collision our present course will bring about.

The environment is suffering critical stress:

The atmosphere Stratospheric ozone depletion threatens us with enhanced ultraviolet radiation at the earth's surface, which can be damaging or lethal to many life forms. Air pollution near ground level, and acid precipitation, are already causing widespread injury to humans, forests, and crops.

Water resources Heedless exploitation of depletable ground water supplies endangers food production and other essential human systems. Heavy demands on the world's surface waters have resulted in serious shortages in some 80 countries, containing 40 percent of the world's population. Pollution of rivers, lakes, and ground water further limits the supply.

Oceans Destructive pressure on the oceans is severe, particularly in the coastal regions which produce most of the world's food fish. The total marine catch is now at or above the estimated maximum sustainable yield. Some fisheries have already shown signs of collapse. Rivers carrying heavy burdens of eroded soil into the seas also carry industrial, municipal, agricultural, and livestock waste—some of it toxic.

Soil Loss of soil productivity, which is causing extensive land abandonment, is a widespread by-product of current practices in agriculture and animal husbandry. Since 1945, 11 percent of the earth's vegetated surface has been degraded—an area larger than India and China combined—and per capita food production in many parts of the world is decreasing.

Forests Tropical rain forests, as well as tropical and temperate dry forests, are being destroyed rapidly. At present rates, some critical forest types will be gone in a few years, and most of the tropical rain forest will be gone before the end of the next century. With them will go large numbers of plant and animal species.

Living species The irreversible loss of species, which by 2100 may reach one-third of all species now living, is especially serious. We are losing the potential they hold for providing medicinal and other benefits, and the contribution that genetic diversity of life forms gives to the robustness of the world's biological systems and to the astonishing beauty of the earth itself.

Much of this damage is irreversible on a scale of centuries, or permanent. Other processes appear to pose additional threats. Increasing levels of gases in the atmosphere from human activities, including carbon dioxide released from fossil fuel burning and from deforestation, may alter climate on a global scale. Predictions of global warming are still uncertain—with projected effects ranging from tolerable to very severe—but the potential risks are very great.

Our massive tampering with the world's interdependent web of life—coupled with the environmental damage inflicted by deforestation, species loss, and climate change—could trigger widespread adverse effects, including unpredictable collapses of critical biological systems whose interactions and dynamics we only imperfectly understand.

Uncertainty over the extent of these effects cannot excuse complacency or delay in facing the threats.

Population The earth is finite. Its ability to absorb wastes and destructive effluent is finite. Its ability to provide food and energy is finite. Its ability to provide for growing numbers of people is finite. And we are fast approaching many of the earth's limits. Current economic practices which damage the environment, in both developed and underdeveloped nations, cannot be continued without the risk that vital global systems will be damaged beyond repair.

Pressures resulting from unrestrained population growth put demands on the natural world that can overwhelm any efforts to achieve a sustainable future. If we are to halt the destruction of our environment, we must accept limits to that growth. A World Bank estimate indicates that world population will not stabilize at less than 12.4 billion, while the United

Nations concludes that the eventual total could reach 14 billion, a near tripling of today's 5.4 billion. But, even at this moment, one person in five lives in absolute poverty without enough to eat, and one in ten suffers serious malnutrition.

No more than one or a few decades remain before the chance to avert the threats we now confront will be lost and the prospects for humanity immeasurably diminished.

Warning

We the undersigned, senior members of the world's scientific community, hereby warn all humanity of what lies ahead. A great change in our stewardship of the earth and the life on it is required, if vast human misery is to be avoided and our global home on this planet is not to be irretrievably mutilated.

What we must do

Six inextricably linked areas must be addressed simultaneously:

1. We must bring environmentally damaging activities under control to restore and protect the integrity of the earth's systems we depend on. We must, for example, move away from fossil fuels to more benign, inexhaustible energy sources to cut greenhouse gas emissions and the pollution of our air and water. Priority must be given to the development of energy sources matched to Third World needs—small-scale and relatively easy to implement.
2. We must halt deforestation, injury to and loss of agricultural land, and the loss of terrestrial and marine plant and animal species.
3. We must manage resources crucial to human welfare more effectively. We must give high priority to efficient use of energy, water, and other materials, including expansion of conservation and recycling.
4. We must stabilize population. This will be possible only if all nations recognize that it requires improved social and economic conditions, and the adoption of effective, voluntary family planning.
5. We must reduce and eventually eliminate poverty.
6. We must ensure sexual equality, and guarantee women control over their own reproductive decisions.

Developed nations must act now

The developed nations are the largest polluters in the world today. They must greatly reduce their overconsumption, if we are to reduce pressures on resources and the global environment. The developed nations have the obligation to provide aid and support to developing nations, because only the developed nations have the financial resources and the technical skills for these tasks.

Acting on this recognition is not altruism, but enlightened self-interest: whether industrialized or not, we all have but one lifeboat. No nation can escape from injury when global biological systems are damaged. No nation can escape from conflicts over increasingly scarce resources. In addition, environmental and economic instabilities will cause mass migrations with incalculable consequences for developed and undeveloped nations alike.

Developing nations must realize that environmental damage is one of the gravest threats they face, and that attempts to blunt it will be overwhelmed if their populations go unchecked. The greatest peril is to become trapped in spirals of environmental decline, poverty, and unrest, leading to social, economic, and environmental collapse.

Success in this global endeavor will require a great reduction in violence and war. Resources now devoted to the preparation and conduct of war—amounting to over $1 trillion annually—will be badly needed in the new tasks and should be diverted to the new challenges.

A new ethic is required—a new attitude towards discharging our responsibility for caring for ourselves and for the earth. We must recognize the earth's limited capacity to provide for us. We must recognize its fragility. We must no longer allow it to be ravaged. This ethic must motivate a great movement, convincing reluctant leaders and reluctant governments and reluctant peoples themselves to affect the needed changes.

The scientists issuing this warning hope that our message will reach and affect people everywhere. We need the help of many.

> We require the help of the world community of scientists—natural, social, economic, and political.
>
> We require the help of the world's business and industrial leaders.

We require the help of the world's religious leaders.

We require the help of the world's peoples.

We call on all to join us in this task.

Caring for the Earth: A Strategy for Sustainable Living

The 1991 publication *Caring for the Earth: A Strategy for Sustainable Living* (Gland, Switzerland: IUCN, UNEP, WWF) offers nine principles that will help individuals, organizations, and nation-states build sustainable societies. Below is a list of these principles, with some of the actions suggested to accomplish each one. (Reprinted with permission.)

Principle 1: **Respect and care for the community of life.**

Actions: Develop a world ethic for sustainable living, promote it at the national level, and implement it through all sectors of society.

Establish a world organization to monitor its implementation and to prevent and combat serious breaches.

Principle 2: **Improve the quality of human life.**

Actions: In lower-income countries, increase economic growth to advance human development.

In upper-income countries, adjust national development policies and strategies to ensure sustainability.

Provide services to promote a long and healthy life.

Provide universal primary education for all children and reduce illiteracy.

Develop more meaningful indicators of quality of life and monitor the extent to which they are achieved.

Enhance security against natural disasters and social strife.

Principle 3: **Conserve the Earth's vitality and diversity.**

Actions:

Adopt a precautionary approach to pollution.

Cut emissions of sulfur dioxide, nitrogen oxides, carbon monoxide, and hydrocarbons.

Reduce greenhouse gas emissions.

Prepare for climate change.

Maintain as much as possible of each country's natural and modified ecosystems.

Take the pressure off natural and modified ecosystems by protecting the best farmland and managing it in ecologically sound ways.

Halt net deforestation, protect large areas of old-growth forest, and maintain a permanent estate of modified forest.

Complete and maintain a comprehensive system of protected areas.

Improve conservation of wild plants and animals.

Improve knowledge and understanding of species and ecosystems.

Use a combination of *in-situ* and *ex-situ* conservation to maintain species and genetic resources.

Harvest wild resources sustainably.

Support management of wild renewable resources by local communities; and increase incentives to conserve biological diversity.

Principle 4: **Minimize the depletion of non-renewable resources.**

Principle 5: **Keep within the Earth's carrying capacity.**

Actions:

Increase awareness about the need to stabilize resource consumption and population, and integrate these issues into national development policies and planning.

Develop, test, and adopt resource-efficient methods and technologies.

Tax energy and other resources in high-consumption countries.

Encourage 'green consumer' movements.

Improve maternal and child health care.

Double family planning services.

Principle 6: **Change personal attitudes and practices.**

Actions:

Ensure that national strategies for sustainability include actions to motivate, educate, and equip individuals to lead sustainable lives.

Review the status of environmental education and make it an integral part of formal education at all levels.

Determine the training needs for a sustainable society and plan to meet them.

Principle 7: **Enable communities to care for their own environments.**

Actions:

Provide communities and individuals with access to resources and an equitable share in managing them.

Improve exchange of information, skills, and technologies.

Enhance participation in conservation and development.

Develop more effective local governments.

Care for the local environment in every community.

Provide financial and technical support to community environmental action.

Principle 8: **Provide a national framework for integrating development and conservation.**

Actions:

Adopt an integrated approach to environmental policy, with sustainability as the overall goal.

Develop strategies for sustainability, and implement them directly and through regional and local planning.

Subject proposed development projects, programs, and policies to environmental impact assessment and to economic appraisal.

Establish a commitment to the principles of a sustainable society in constitutional or other fundamental statements of national policy.

Establish a comprehensive system of environmental law and provide for its implementation and enforcement.

Review the adequacy of legal and administrative controls, and of implementation and enforcement mechanisms, recognizing the legitimacy of local approaches.

Ensure that national policies, development plans, budgets, and decisions on investments take full account of their effects on the environment.

Use economic policies to achieve sustainability.

Provide economic incentives for conservation and sustainable use.

Strengthen the knowledge base, and make information on environmental matters more accessible.

Principle 9: **Create a global alliance.**

Actions: Strengthen existing international agreements and conclude new ones to conserve life-support systems and biological diversity.

Develop a comprehensive and integrated conservation regime for Antarctica and the Southern Ocean.

Prepare and adopt a Universal Declaration and Covenant on Sustainability.

Write off the official debt of low-income countries, and retire enough of their commercial debt to restore economic progress.

Increase the capacity of lower-income countries to help themselves.

Increase development assistance and devote it to helping countries develop sustainable societies and economies.

Recognize the value of global and national non-governmental action, and strengthen it.

Strengthen the United Nations system as an effective force for global sustainability.

National Biodiversity Institute Documents

Costa Rica's National Biodiversity Institute (INBio) was founded in 1989 and serves as a model for other biodiverse countries whose governments are beginning to see their biodiversity as a valuable national resource. Following are two INBio documents: INBio's Mission Statement, and the contract it signed with the giant pharmaceutical firm Merck & Company, Inc. (Reprinted with permission of Instituto Nacional de Biodiversidad.)

Mission Statement of Costa Rica's National Biodiversity Institute

Costa Rica's National Biodiversity Institute (INBio) is a private, non-profit scientific institution with a social orientation and for the public good. Its mission is to promote greater awareness of the value of biodiversity, and thereby achieve its conservation and improve the quality of life for society.

INBio generates knowledge about biodiversity. It disseminates and promotes this information in diverse targeted formats designed to be responsive to a broad spectrum of national and international users. Its activities support the spiritual, social and economic development of Costa Rican society in harmony with the environment.

The mission is carried out through the integration of the following processes:

- A systematic biodiversity inventory particularly of Costa Rica's protected areas;
- The quest for sustainable uses of biodiversity for and by all sectors of society, and promotion of these uses;
- Organization and management of biodiversity information;

- Generation and dissemination of biodiversity knowledge.

INBio acts with a humanistic vision that is innovative, organized, participatory and multidisciplinary, and through strategic alliances with many national and international sectors.

Summary of Terms for the INBio-Merck & Co., Inc. Collaboration Agreement

Parties: Asociación Instituto Nacional de Biodiversidad, a non-profit organization existing under the laws of Costa Rica. ("INBio") / Merck & Co., Inc., a corporation organized under the laws of the State of New Jersey, U.S.A. ("Merck").

Effective Date: Agreement signed November 1, 1991/ Renewed July 1994 and August 1996 for two additional years each time.

Purpose of Agreement: INBio is interested in collaborating with private industry to create mechanisms to help preserve Costa Rican conservation areas by making them economically viable.

Merck is interested in collaborating with INBio to obtain plant, insect and environmental samples for evaluation for pharmaceutical and agricultural applications.

Obligations of INBio: INBio agrees to establish facilities for the collection and processing of plant, insect and environmental samples from Costa Rica.

INBio agrees to hire and train an adequate staff to collect and process the samples.

Merck agrees to provide training to INBio staff in Merck facilities.

INBio agrees to supply Merck with a specified number of plant, insect and environmental samples per year over the initial two-year period of the Agreement as described in the workplan.

The plant and insect samples will be processed in laboratory facilities which were initially established by INBio at the University of Costa Rica, and are currently being established at INBio.

Obligations of Merck: Merck agrees to provide research funding of $1.0 million during the first two years of the Agreement and to contribute to INBio laboratory equipment and materials needed to operate the processing laboratory.

Merck agrees to evaluate the samples provided by INBio in proprietary assays for potential activity as human health, animal health, and agricultural compounds. Merck agrees to advise INBio of confirmed and reproducible activity that has been identified in any INBio samples.

Merck agrees to assign unique identification numbers to all INBio samples and to maintain an identification system which will allow Merck and INBio to identify all products which may be subject to royalty under the Agreement.

Exclusivity of Arrangement: INBio agrees that during an initial evaluation period of two years it will not provide any samples that have been provided to Merck to other parties for use in the field of human and animal health and agriculture. However, INBio may provide the Merck samples to parties for evaluation outside of Merck's field of interest.

After the initial evaluation period is completed, INBio will be free to offer the samples that have been supplied to Merck to other parties for evaluation for human health, animal health or agricultural uses. With respect to no more than 1% of the total number of samples provided by INBio to Merck, Merck may extend the exclusive evaluation period as long as Merck acts diligently in the evaluation and commercialization of the sample. The exclusive rights will terminate if Merck ceases the program to commercialize products derived from the sample.

INBio may decline to obtain a sample if the samples are impossible to obtain for reasons related to logistics, endangered species concerns, or biological concerns.

Merck will provide INBio with written progress reports at least once each year concerning its commercialization activities with respect to a specific sample.

Confidentiality: During the term of the Agreement and for a seven year period thereafter, the parties agree not to disclose any confidential information received from the other party under the collaboration to any third party.

Either party may publish the results of the research collaboration after providing the other party with the opportunity to review the publication.

Invention and Patents: Inventions made as part of the research collaboration will be owned by Merck, and Merck will be responsible for filing appropriate patent applications. INBio will be compensated for its contribution to any invention by a royalty on sales of products, as described in Payments. INBio retains the right to provide samples to third parties for evaluation and commercial development, subject to the limited exclusivity granted to Merck.

Payments: Merck agrees to pay a royalty to INBio on any human or animal pharmaceutical product or agricultural chemical compound which is isolated initially from or produced by a sample INBio provided to Merck. The royalty obligation also applies to any products which are derivatives or analogs of such compounds. The royalty obligation applies to chemical compounds derived either from living isolates from environmental samples, or from samples of dead tissues.

The royalty rate is confidential business information and will not be disclosed. The royalty rate falls into the range of royalty rates typical for agreements of this kind.

Merck agrees to maintain accurate records which will allow Merck and INBio to identify all products subject to royalty and to enable INBio to confirm the accuracy of Merck's royalty reports.

Indemnification: Merck agrees to indemnify INBio from any claims arising from the use of the samples, except for any claims resulting from the negligence or other wrongful act of INBio.

Merck agrees to comply with all regulatory and other requirements which apply to the use of the samples.

Term: The initial term of the Agreement will be two years from the date on which the processing facilities are available for operation.

Three months prior to the termination of the initial term, or any extension after the initial term, the parties shall meet to determine whether to extend the collaboration for an additional period. Merck will provide additional funding in an agreed amount to support INBio's work during any extension period.

Termination: Either party may terminate on ninety days' written notice in the event of a material breach of the contract by the other party.

Either party may terminate if the other party becomes insolvent, makes an assignment for the benefit of creditors, or is the subject of bankruptcy proceedings.

In the event of termination, the confidential obligations and royalty obligations shall remain in effect.

Assignments and sublicensing: Neither party may assign the Agreement.

Merck may enter into sublicensing Agreements provided that Merck remains liable to INBio for any obligations under the Agreement and that all royalties due to INBio are paid. Merck shall notify INBio of any sublicense which involves the license of INBio samples or confidential INBio information.

The Convention on Biological Diversity (1992)

The United Nations Environmental Programme sponsored an "Earth Summit" in June 1992, which was held in Rio de Janeiro, Brazil. It resulted in Agenda 21, a series of conventions including one on climate change, another on environment and development, the "Forest Principles," and the Convention on Biodiversity. The Convention on Biodiversity had been signed and ratified by 169 countries as of June 1997. The Convention on Biological Diversity demonstrates a strong commitment to the needs of biologically rich developing countries, which will not be able to protect their biodiversity without significant financial and technical assistance from their wealthier counterparts. Reprinted below is the "meat" of the Convention, specifically the Preamble, which introduces the motivations behind the Convention, and some of the responsibilities for conservation taken on by the signees. The Convention in its entirety can be found in government documents sections of libraries or on many World Wide Web sites.

Preamble

The Contracting Parties,

Conscious of the intrinsic value of biological diversity and of the ecological, genetic, social, economic, scientific, educational, cultural, recreational and aesthetic values of biological diversity and its components,

Conscious also of the importance of biological diversity for evolution and for maintaining life sustaining systems of the biosphere,

Affirming that the conservation of biological diversity is a common concern of humankind,

Reaffirming that States have sovereign rights over their own biological resources,

Reaffirming also that States are responsible for conserving their biological diversity and for using their biological resources in a sustainable manner,

Concerned that biological diversity is being significantly reduced by certain human activities,

Aware of the general lack of information and knowledge regarding biological diversity and of the urgent need to develop scientific, technical and institutional capacities to provide the basic understanding upon which to plan and implement appropriate measures,

Noting that it is vital to anticipate, prevent and attack the causes of significant reduction or loss of biological diversity at source,

Noting also that where there is a threat of significant reduction or loss of biological diversity, lack of full scientific certainty should not be used as a reason for postponing measures to avoid or minimize such a threat,

Noting further that the fundamental requirement for the conservation of biological diversity is the *in-situ* conservation of ecosystems and natural habitats and the maintenance and recovery of viable populations of species in their natural surroundings,

Noting further that *ex-situ* measures, preferably in the country of origin, also have an important role to play,

Recognizing the close and traditional dependence of many indigenous and local communities embodying traditional lifestyles on biological resources, and the desirability of sharing equitably benefits arising from the use of traditional knowledge, innovations and practices relevant to the conservation of biological diversity and the sustainable use of its components,

Recognizing also the vital role that women play in the conservation and sustainable use of biological diversity and affirming the need for the full participation of women at all levels of policy-making and implementation for biological diversity conservation,

Stressing the importance of, and the need to promote, international, regional and global cooperation among States and intergovernmental organizations and the non-governmental sector for the conservation of biological diversity and the sustainable use of its components,

Acknowledging that the provision of new and additional financial resources and appropriate access to relevant technologies can be expected to make a substantial difference in the world's ability to address the loss of biological diversity,

Acknowledging further that special provision is required to meet the needs of developing countries, including the provision of new and additional financial resources and appropriate access to relevant technologies,

Noting in this regard the special conditions of the least developed countries and small island States,

Acknowledging that substantial investments are required to conserve biological diversity and that there is the expectation of a broad range of environmental, economic and social benefits from those investments,

Recognizing that economic and social development and poverty eradication are the first and overriding priorities of developing countries,

Aware that conservation and sustainable use of biological diversity is of critical importance for meeting the food, health and other needs of the growing world population, for which purpose access to and sharing of both genetic resources and technologies are essential,

Noting that, ultimately, the conservation and sustainable use of biological diversity will strengthen friendly relations among States and contribute to peace for humankind,

Desiring to enhance and complement existing international arrangements for the conservation of biological diversity and sustainable use of its components, and

Determined to conserve and sustainably use biological diversity for the benefit of present and future generations,

Have agreed as follows:

Article 1. Objectives

The objectives of this Convention, to be pursued in accordance with its relevant provisions, are the conservation of biological diversity, the sustainable use of its components and the fair and equitable sharing of the benefits arising out of the utilization of genetic resources, including by appropriate access to genetic resources and by appropriate transfer of relevant technologies, taking into account all rights over those resources and to technologies, and by appropriate funding.

Article 6. General Measures for Conservation and Sustainable Use

Each Contracting Party shall, in accordance with its particular conditions and capabilities:

(a) Develop national strategies, plans or programmes for the conservation and sustainable use of biological diversity or adapt for this purpose existing strategies, plans or programmes which shall reflect, *inter alia*, the measures set out in this Convention relevant to the Contracting Party concerned; and

(b) Integrate, as far as possible and as appropriate, the conservation and sustainable use of biological diversity into relevant sectoral or cross-sectoral plans, programmes and policies.

Article 7. Identification and Monitoring

Each Contracting Party shall, as far as possible and as appropriate, in particular for the purposes of Articles 8 to 10:

(a) Identify components of biological diversity important for its conservation and sustainable use having regard to the indicative list of categories set down in Annex I;

(b) Monitor, through sampling and other techniques, the components of biological diversity identified pursuant to subparagraph (a) above, paying particular attention to those requiring urgent conservation measures and those which offer the greatest potential for sustainable use;

(c) Identify processes and categories of activities which have or are likely to have significant adverse impacts on the conservation and sustainable use of biological diversity, and monitor their effects through sampling and other techniques; and

(d) Maintain and organize, by any mechanism data, derived from identification and monitoring activities pursuant to subparagraphs (a), (b) and (c) above.

Article 8. In-situ Conservation

Each Contracting Party shall, as far as possible and as appropriate:

(a) Establish a system of protected areas or areas where special measures need to be taken to conserve biological diversity;

(b) Develop, where necessary, guidelines for the selection, establishment and management of protected areas or areas where special measures need to be taken to conserve biological diversity;

(c) Regulate or manage biological resources important for the conservation of biological diversity whether within or outside protected areas, with a view to ensuring their conservation and sustainable use;

(d) Promote the protection of ecosystems, natural habitats and the maintenance of viable populations of species in natural surroundings;

(e) Promote environmentally sound and sustainable development in areas adjacent to protected areas with a view to furthering protection of these areas;

(f) Rehabilitate and restore degraded ecosystems and promote the recovery of threatened species, inter alia, through the development and implementation of plans or other management strategies;

(g) Establish or maintain means to regulate, manage or control the risks associated with the use and release of living modified organisms resulting from biotechnology which are likely to have adverse environmental impacts that could affect the conservation and sustainable use of biological diversity, taking also into account the risks to human health;

(h) Prevent the introduction of, control or eradicate those alien species which threaten ecosystems, habitats or species;

(i) Endeavour to provide the conditions needed for compatibility between present uses and the conservation of biological diversity and the sustainable use of its components;

(j) Subject to its national legislation, respect, preserve and maintain knowledge, innovations and practices of indigenous and local communities embodying traditional lifestyles relevant for the conservation and sustainable use of biological diversity and promote their wider application with the approval and involvement of the holders of such knowledge, innovations and practices and encourage the equitable sharing of the benefits arising from the utilization of such knowledge, innovations and practices;

(k) Develop or maintain necessary legislation and/or other regulatory provisions for the protection of threatened species and populations;

(l) Where a significant adverse effect on biological diversity has been determined pursuant to Article 7, regulate or manage the relevant processes and categories of activities; and

(m) Cooperate in providing financial and other support for *in-situ* conservation outlined in subparagraphs (a) to (l) above, particularly to developing countries.

Article 9. Ex-situ Conservation

Each Contracting Party shall, as far as possible and as appropriate, and predominantly for the purpose of complementing *in-situ* measures:

(a) Adopt measures for the *ex-situ* conservation of components of biological diversity, preferably in the country of origin of such components;

(b) Establish and maintain facilities for *ex-situ* conservation of and research on plants, animals and micro-organisms, preferably in the country of origin of genetic resources;

(c) Adopt measures for the recovery and rehabilitation of threatened species and for their reintroduction into their natural habitats under appropriate conditions;

(d) Regulate and manage collection of biological resources from natural habitats for *ex-situ* conservation purposes so as not to threaten ecosystems and *in-situ* populations of species, except where special temporary *ex-situ* measures are required under subparagraph (c) above; and

(e) Cooperate in providing financial and other support for *ex-situ* conservation outlined in subparagraphs (a) to (d) above and in the establishment and maintenance of *ex-situ* conservation facilities in developing countries.

Article 10. Sustainable Use of Components of Biological Diversity

Each Contracting Party shall, as far as possible and as appropriate:

(a) Integrate consideration of the conservation and sustainable use of biological resources into national decision-making;

(b) Adopt measures relating to the use of biological resources to avoid or minimize adverse impacts on biological diversity;

(c) Protect and encourage customary use of biological resources in accordance with traditional cultural practices that are compatible with conservation or sustainable use requirements;

(d) Support local populations to develop and implement remedial action in degraded areas where biological diversity has been reduced; and

(e) Encourage cooperation between its governmental authorities and its private sector in developing methods for sustainable use of biological resources.

Article 11. Incentive Measures

Each Contracting Party shall, as far as possible and as appropriate, adopt economically and socially sound measures that act as incentives for the conservation and sustainable use of components of biological diversity.

Article 12. Research and Training

The Contracting Parties, taking into account the special needs of developing countries, shall:

(a) Establish and maintain programmes for scientific and technical education and training in measures for the identification, conservation and sustainable use of biological diversity and its components and provide support for such education and training for the specific needs of developing countries;

(b) Promote and encourage research which contributes to the conservation and sustainable use of biological diversity, particularly in developing countries, *inter alia*, in accordance with decisions of the Conference of the Parties taken in consequence of recommendations of the Subsidiary Body on Scientific, Technical and Technological Advice; and

(c) In keeping with the provisions of Articles 16, 18 and 20, promote and cooperate in the use of scientific advances in biological diversity research in developing methods for conservation and sustainable use of biological resources.

Article 13. Public Education and Awareness

The Contracting Parties shall:

(a) Promote and encourage understanding of the importance of, and the measures required for, the conservation of biological diversity, as well as its propagation through media, and the inclusion of these topics in educational programmes; and

(b) Cooperate, as appropriate, with other States and international organizations in developing educational and public awareness programmes, with respect to conservation and sustainable use of biological diversity.

Article 14. Impact Assessment and Minimizing Adverse Impacts

1. Each Contracting Party, as far as possible and as appropriate, shall:

    (a) Introduce appropriate procedures requiring environmental impact assessment of its proposed projects that are likely to have significant adverse effects on biological diversity with a view to avoiding or minimizing such effects and, where appropriate, allow for public participation in such procedures;

    (b) Introduce appropriate arrangements to ensure that the environmental consequences of its programmes and policies that are likely to have significant adverse impacts on biological diversity are duly taken into account;

    (c) Promote, on the basis of reciprocity, notification, exchange of information and consultation on activities under their jurisdiction or control which are likely to significantly affect adversely the biological diversity of other States or areas beyond the limits of national jurisdiction, by encouraging the conclusion of bilateral, regional or multilateral arrangements, as appropriate;

    (d) In the case of imminent or grave danger or damage, originating under its jurisdiction or control, to biological diversity within the area under jurisdiction of other States or in areas beyond the limits of national jurisdiction, notify immediately the potentially affected States of such danger or damage, as well as initiate action to prevent or minimize such danger or damage; and

    (e) Promote national arrangements for emergency responses to activities or events, whether caused naturally or otherwise, which present a grave and imminent danger to biological diversity and encourage international cooperation to supplement such national efforts and, where appropriate and agreed by the States or regional economic integration organizations concerned, to establish joint contingency plans.

2. The Conference of the Parties shall examine, on the basis of studies to be carried out, the issue of liability and redress, including restoration and compensation, for damage to biological diversity, except where such liability is a purely internal matter.

Directory of Organizations

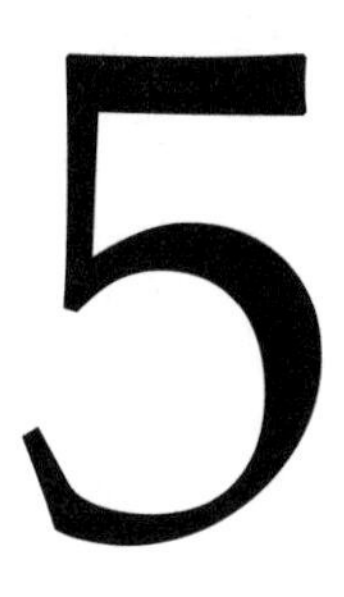

Many thousands of organizations worldwide are working on issues related to biodiversity. Since biodiversity is a multifaceted topic, it is included in the agendas of organizations devoted to biological study, environmental conservation, sustainable agriculture and development, and the rights of indigenous peoples.

This chapter lists and describes some of the organizations that study and document biodiversity, and work to protect it. Many more organizations can be found in the annual conservation directory published by the National Wildlife Federation, or by searching the World Wide Web. Brief descriptions of some conservation organizations, along with links to their web sites, are given by EnviroLink (http://envirolink.org/) and the Wilderness Society (http://www.lbbs.org/wild/about.htm).

American Littoral Society
Sandy Hook
Highlands, NJ 07732
(908) 291-0055
Fax: (908) 872-8041

The word *littoral* refers to coastal or shore regions, and this organization is concerned with littoral ecosystems like tidal rivers, estuaries, and coastal wetlands. Humans tend to cluster in littoral zones (three-quarters of the world's human population lives in coastal regions), and tempering our impact on these nutrient-rich ecosystems is a challenge. This organization contributes by organizing citizens to monitor coastal water quality and report polluters, leading a volunteer fish tag-and-release program that provides data on migration and population rates, protecting public access to beaches, and lobbying state and federal representatives on coastal protection issues. Field trips varying from evening walks to weeklong expeditions are offered to members.

Publications: Underwater Naturalist is a quarterly journal sent to all members. Members also receive the *Coastal Reporter*, a national newsletter, and several regional chapters publish newsletters, too.

American Livestock Breeds Conservancy (ALBC)
P.O. Box 477
Pittsboro, NC 27312
(919) 542-5704
Fax: (919) 545-0022

Modern animal husbandry has essentially bred itself into a corner. Industry breeders select for the most favored attribute in farm animals: high productivity. However, other features of livestock are important as well: fertility, foraging ability, longevity, strong maternal instincts, and resistance to disease and parasites. Yet many of these qualities are in danger of disappearing from common breeds as their genetic diversity erodes. "Heritage breeds," those varieties of livestock that were once popular but are now dying out, do exhibit these and other important attributes and should be protected so that they can contribute genetic diversity to the more common breeds. ALBC serves as a clearinghouse for livestock and genetic diversity and seeks to protect almost 100 domestic breeds from extinction. It maintains gene banks, rescues threatened populations, conducts research on breeds, educates on genetic diversity of livestock and the role of livestock in sustainable agriculture, and supports breeders and farmers.

Publications: ALBC News is a bimonthly newsletter for members. ALBC offers a catalog of publications mostly aimed at farmers and their children, with some books oriented toward a general audience.

Arbofilia
Apartado 512-1100
Tibás, San José
Costa Rica
Phone/Fax: (506) 240-7145

This Costa Rican organization replenishes lost biodiversity by helping regenerate the ecology of deforested watersheds in the most devastated area of Costa Rica. Local *campesinos* (small-scale family farmers) raise seedlings of native tree species, then plant them along dried-up streambeds. As the trees mature and shade the springs and streams, the water returns. Arbofilia is also a proponent of agricultural alternatives to monocropping. It promotes "analog forests," which layer several different crops growing at different heights: ground cover, bushes, vines, and trees, for example. Through this type of agriculture, a habitat for animal species is created and the soil's health is conserved at the same time that the land can produce a variety of staples and cash crops.

Bat Conservation International (BCI)
P.O. Box 162603
Austin, TX 78716
(512) 327-9721
Fax: (512) 327-9724
E-mail: batinfo@batcon.org
Web site: http://www.batcon.org/

The roles of bats in pollination and insect control is often underappreciated, but BCI is an effective public relations organization for these nocturnal mammals. Fourteen thousand members in 71 countries have joined BCI to work on bats' behalf. BCI prides itself not on how many conservation battles it has won but rather how many it has been able to avoid. It converts enemies into friends through its innovative alliances with highway department engineers and miners. These two professions can play an important role for bat conservation because as their natural habitat shrinks, bats are roosting more frequently under bridges and in mining shafts. Very simple changes in bridge design and mine gating can result in homey roosts for bats. Farmers are also learning how to build bat houses on their property, which take the place of disappearing forest snags. BCI designed the Masters of the Night museum exhibit, which has toured natural history and science museums throughout the country, and BCI holds bat conservation training workshops internationally.

Publication: Bats is a quarterly magazine for members.

Biodiversity Legal Foundation (BLF)
P.O. Box 18327
Boulder, CO 80308-1327
E-mail: blfrog@aol.com

This foundation works with scientists and grassroots conservation groups in two specific ways: by gathering data on endangered species and working with the U.S. Fish and Wildlife Service (USFWS) to assure their protection. BLF identifies rare and imperiled species in North America and prepares biological reviews of their status. This documentation, along with a formal petition for placement on the endangered species list, is presented to the USFWS (see separate listing in this chapter), the federal agency charged with maintaining the endangered species list. BLF helps develop plans for management and recovery of these and other listed species and their ecosystems and monitors the USFWS's compliance with its mandate to protect endangered species and their habitats. When the USFWS fails, BLF is prepared to take administrative or legal actions against it.

Birdlife International
Wellbrook Court
Girton Road
Cambridge CB3 ONA
United Kingdom
Phone: 01223-277318
Fax: 01223-277200
E-mail: birdlife@gn.apc.org

This organization, founded in 1922 and currently active in 100 countries, works to conserve bird species and their habitats. Birds are good indicators of biological richness and environmental trends, and they are effective ambassadors for conservation because humans take such delight in them. The organization recognizes that the problems confronting birds are inextricably linked to social, economic, and cultural factors, and that the only solution lies in revamping of society to function in an ecologically sustainable manner. Birdlife International performs scientific research and analysis, helps develop environmental policy, participates in community-based land-use and management projects, and builds networks.

Publications: Members receive a quarterly magazine. The Birdlife Conservation Series are books listing and describing threatened species all over the globe, used by IUCN (see below) as a source for its Red Data Books.

Center for Marine Conservation (CMC)
1725 DeSales Street NW, #600
Washington, DC 20036
(202) 429-5609
Fax: (202) 872-0619
Web site: http://www.cmc-ocean.org

Oceans and seas cover three-quarters of the planet, provide over half of the animal protein in the diet of many peoples, produce a third of the oxygen in our air, and harbor potentially helpful medicines for humanity. Vast as they are, however, they are not infinite in their ability to provide food or to absorb the waste dumped into them. CMC works for conservation of all forms of marine life and marine ecosystems in general. An Ocean Action Network asks members to write letters on urgent situations. An annual coastal cleanup in September mobilizes volunteers all over the world and serves as an opportunity to educate the public about the importance of the planet's seas. CMC classifies the debris collected and publicizes information about its components. CMC is especially concerned with plastic that finds its way into the ocean, since it can prove deadly to wildlife.

Publications: A variety of kits are available about different types of pollution affecting marine ecosystems: balloons, plastic, contamination from recreational and commercial fishing, and cruise boats. Many other publications are listed on CMC's web site.

Center for Plant Conservation (CPC)
Missouri Botanical Garden
P.O. Box 299
St. Louis, MO 63166-0299
(314) 577-9450
E-mail: CPC@mobot.org
Web site: http://www.mobot.org/cpc

Of the 20,000 plant species native to the United States, more than one-fifth are threatened to some extent. The Center for Plant Conservation, founded in 1984, is the only national organization dedicated exclusively to the preservation of the country's endangered flora. With headquarters at the Missouri Botanical Garden, the CPC is a network of 28 institutions throughout the continental United States and Hawaii. Each member works with plants of particular concern in its region. The CPC is particularly active in Hawaii, where there is a high percentage of endemic plants. Between its members, the CPC maintains almost 500 endangered plant species. Plants (preferably in seed form) are gathered in the wild and kept in their seed, cutting, and mature forms. CPC member institutions perform research on these species, educate the public about them, and provide information to other scientists and conservationists. Their goal is to maintain the genetic diversity of these species until a viable wild population can be restored. Along with its *ex-situ* projects, the CPC has pioneered an *in-situ* technique called the integrated conservation methodology, which involves collaboration from participating institutions, researchers, and land management organizations, each contributing its own resources and skills to a successful conservation project.

Publications: Members receive CPC's newsletter, *Plant Conservation*. CPC publishes rare plant directories and conservation guides and some technical books. *Plants in Peril* ($3) is an attractive and useful 24-page booklet for middle school educators, including basic information on plant biodiversity and threats to it, and some great classroom activities.

Conservation International (CI)
2501 M Street NW, #200
Washington, DC 20037
(202) 429-5660
Web site: http://www.conservation.org/

Focusing on the 22 biologically richest countries of the world, CI works toward the two-part goal of environmental conservation and improved quality of life for human residents. Its five strategies include increasing scientific understanding of biodiversity, developing economic alternatives to environmentally destructive activities, assisting in the design and implementation of conservation policies, maintaining parks or protected areas, and building an awareness, commitment, and capacity for conservation in local communities, governments, and the private sector. Some of CI's current projects include a Shaman's Apprentice Program in Suriname, in which young indigenous people learn about traditional medicine and healing plants from their elderly shamans; facilitating fair bioprospecting agreements between the government of Suriname, a major international pharmaceutical firm, and the U.S. National Institute of Health; helping the governments of Madagascar and Indonesia write legislation protecting their rights to the genetic resources in their forests; and promoting the sustainable harvest of a rain forest seed called tagua, which is as beautiful and hard as ivory and can be carved into jewelry and buttons.

Publications: CI has produced several videos on biodiverse regions of the world.

Defenders of Wildlife
1101 14th Street NW, #1400
Washington, DC 20005
(202) 682-9400
E-mail: webmaster@defenders.org
Web site: http://www.defenders.org/

Defenders of Wildlife focuses on stemming the planet's loss of biodiversity. It is active in protecting entire ecosystems and landscapes and in monitoring indicator species—those whose disappearance could signal the ecological breakdown of their habitat. Defenders has traditionally put special attention into protecting predators. Since large predators have wide ranges, protecting their entire habitat means more biodiversity will be saved. Predators also seem to maintain the balance of many other species in an ecosystem. Defenders led an innovative fight to restore the gray wolf to the Yellowstone area, with special considerations for local ranchers who feared that the wolves would threaten their livestock. For example, their Wolf Compensation Trust reimburses ranchers at fair market values for verified losses of livestock to wolves, and another program involves local residents in the program by paying them to allow wolf pups to be born and raised on their property. Defenders is currently working to restore other major predators: black bears in Florida, grizzlies in Idaho and Montana, and polar bears in Alaska. Defenders was founded in 1947 and has worked on much of the conservation legislation we now take for granted, including the Endangered Species Act and the Convention on International Trade in Endangered Species of Wild Fauna and Flora.

Publications: Defenders magazine is published quarterly. Defenders' web site has a state-by-state biodiversity conservation report.

Desert Fishes Council
P.O. Box 337
Bishop, CA 93515-0337
Phone/Fax: (619) 872-8751
E-mail: phildesfish@telis.org
Web site: http://www.utexas.edu/depts/tnhc/.www/fish/def/dfc_top.html

Because desert fishes have evolved in response to unusual conditions—very small pools and streams where temperatures exceed 120 degrees Fahrenheit in summer and where the level of salinity is five times greater than that of the ocean—they have attracted the interest and protective instincts of scientists and environmentalists. Water is an especially precious resource in the desert, so it is not surprising that what little there is is in great demand by an ever-expanding human population. When water is pumped from the ground, natural springs dry up and the fish lose their homes. Of the 83 fishes listed as endangered or threatened under the Endangered Species Act, 50 are desert fishes. The Desert Fishes Council formed in 1969 to address degradation of aquatic habitats in the southwestern United States and northern Mexico. Its original concerns were problems in Death Valley and the adjacent Ash Meadows. The Desert Fishes Council worked with the Nature Conservancy to buy out developers who were causing the problems, and later the U.S. Fish and Wildlife Service stepped in to formally protect the area. The council has more recently expanded its attention to cover Colorado River ecosystems.

The E. F. Schumacher Society
140 Jug End Road
Great Barrington, MA 01230
(413) 528-1737
E-mail: efssociety@aol.com
Web site: http://www.schumachersociety.org

The E. F. Schumacher Society was founded in 1981 to promote the ideas of its namesake, the author of the 1973 classic *Small Is Beautiful: Economics as If People Mattered*. The book linked economics, ecology, and culture, and was a precursor to today's widely discussed "sustainable development." The society promotes Schumacher's ideas in several ways. It sponsors an annual lecture series about issues such as land use, agriculture, community, and the balance between human needs and the well-being of the natural world. It advises individuals interested in community land trusts, a method of taking land out of the real estate market and assuring that it is used for affordable housing, sustainable agriculture, or open space. One of the group's current projects is the promotion of "local currency," an economic system wherein community members issue their own currency, which can be used within the local economy—one step beyond bartering. Their center is located on a land trust site and includes a library of related literature and E. F. Schumacher's personal collection of books.

Publications: Local Currency News, the society's newsletter, covers community experiments in local currency. Transcripts of more than 40 lectures are available in pamphlet format from the society.

Earth First!
P.O. Box 1415
Eugene, OR 97740
(503) 741-9191
Fax: (503) 741-9192
E-mail: earthfirst@igc.apc.org
Web site: http://www.telalink.net/~zoomst/earthfirst/efinfo.html

At a time when environmentalism is becoming accepted by mainstream society and environmentalist organizations are hammering out compromises with industry and policymakers, Earth First! maintains a policy of "No Compromise for Mother Earth!" EF! is a movement rather than an organization, but a journal (see below) networks Earth First!ers. Earth First!ers put themselves on the line for their beliefs by taking direct action for wilderness protection. Their activities range from grassroots organizing to litigation to civil disobedience to "monkeywrenching" (the act of sabotaging heavy machinery used in environmentally destructive work—given its name and popularity by Edward Abbey's *The Monkey Wrench Gang*). Examples of EF! actions include blocking logging roads and driving spikes into trees so that loggers won't cut them and risk being injured when their saws hit the stakes. Because their radical actions often receive media coverage, Earth First! succeeds in bringing public attention to environmental issues that might otherwise go unnoticed.

Publications: Earth First! Journal is published eight times a year on pagan holidays. It runs essays on preservation of biodiversity and wilderness and chronicles actions of Earth First! and the radical environmental movement.

Earth Island Institute (EII)
300 Broadway, #28
San Francisco, CA 94133-3122
(415) 788-3666
Fax: (415) 788-7324
E-mail: earthisland@earthisland.org
Web site: http://www.earthisland.org/

Earth Island Institute provides an administrative framework to a variety of very specific conservation and sustainable development– oriented projects throughout the world. Once EII adopts a project, it will offer technical assistance, fund-raising support, and possibilities for networking. Some examples of projects it supports include: protection and restoration of mangroves; conservation of the biodiversity of the Tibetan Plateau, especially sources of medicinal plants; protection of sea turtle nesting sites; eliminating two-stroke boat motors, which according to project literature collectively dump more fuel into waterways than the Exxon *Valdez* spill did; and support of the efforts of the Uma Bawang village in Malaysian Borneo, which upholds traditional sustainable lifestyles in the face of massive logging of their rain forest.

Publication: Earth Island Journal is published quarterly.

EarthSave International
706 Frederick Street
Santa Cruz, CA 95062-2206
(408) 423-4069
Web site: http://www.earthsave.org

This organization, founded in 1989 by Baskin-Robbins heir John Robbins in response to a surge of interest generated by his book *Diet for a New America*, illuminates the ties between human and environmental health. Our voracious appetite for meat has devastating consequences both for us personally (70 percent of disease in the United States is diet-related) and for the environment (industrial-scale meat production results in overuse and pollution of water, overconsumption of energy, erosion and loss of topsoil, use of toxic pesticides and chemical fertilizers, depletion of fish from the oceans, and loss of natural habitat—all factors leading to the loss of biodiversity). EarthSave members form Local EarthSave Groups (LEGs) to bring these issues into their communities; kids are empowered by YES! (Youth for Environmental Sanity), a roving multimedia performance by young people; and their Healthy School Lunch program educates school lunch personnel and students about plant-based meals.

Publications: Members receive a subscription to EarthSave's quarterly *NewsMagazine* and *Realities for the '90s*, a booklet with facts about health-environment links. EarthSave's catalog offers a variety of videos, audiotapes, posters, and books.

Earthwatch
P.O. Box 9104
Watertown, MA 02272-9104
(800) 776-0188
Fax: (617) 926-8532
E-mail: info@earthwatch.org
Web site: http://www.earthwatch.org

Earthwatch invites volunteers to participate in more than 120 scientific expeditions each year to far reaches of every continent on earth. Although they are not limited to ecology, many of the trips focus on biodiversity: Puerto Rican Songbirds, Owl Monkeys of Argentina, Forest of the Black Lemur in Madagascar, and Saving Philippine Reefs, for example. The expeditions are led by scientists from the United States or the host countries. Volunteers range from high school students to truck drivers to chief executive officers of large corporations.

Publication: An annual expedition guide describes all upcoming trips.

EnviroLink
Web site: http://www.envirolink.org/

This nonprofit organization connects environmental groups via technology like the Internet. Besides providing links to hundreds of organizations and summaries of their activities, EnviroLink runs specific projects: EcoLex, easy access to environmental laws; an on-line resource guide to the Endangered Species Act; an EnviroArts Gallery; EnviroChat and EnviroForum, two on-line discussion areas; EnviroLink Library, a list of organizations, publications, and government agencies pertinent to environmental work; an environmental education network; mailing lists; and a world species list.

Friends of the Earth (FOE)
1025 Vermont Avenue NW, #300
Washington, DC 20005-6303
(202) 783-7400
Fax: (202) 783-0444
Web site: http://www.foe.org/

Founded in 1969, Friends of the Earth has been working to clean up the basic elements of our natural environment: air, water, ecosystems, and the atmosphere. It provides support to grassroots activists working on these projects at local levels; at the national and international levels it runs two programs. The first, Economics for the Earth, works to eliminate environmentally destructive government subsidies and replace them with sustainable development and economic security initiatives. This program publishes an annual exposé on these subsidies (see below), campaigns for a reformed tax system, monitors national budgets and spending, and works against excessive roadway construction, which is both costly and environmentally damaging. The second project, Global Action, collaborates with more than 50 Friends of the Earth groups worldwide on global issues such as World Bank and International Monetary Fund environmental policies, ozone layer protection, and sustainable world tourism.

Publications: An annual *Green Scissors* report exposes wasteful government spending on environmentally harmful projects. *Road to Ruin* describes 22 unnecessary road construction projects. Other papers and books are also available and are listed in FOE's web site.

Global Response (GR)
Environmental Action Network
P.O. Box 7490
Boulder, CO 80306-7490
(303) 444-0306
Fax: (303) 449-9794
E-mail: globresponse@igc.apc.org
Web site: http://www.globalresponse.org

Global Response describes itself as "an international letter-writing network of dedicated global citizens who respond to specific environmental threats by mobilizing broad-based campaigns to bring about positive change." Global Response issues monthly Actions, which are developed in partnership with local grassroots indigenous, environmentalist, and peace and justice groups around the world. They address problems such as environmental destruction, atmospheric and marine contamination, toxic waste dumping, threats to indigenous peoples, and threats to marine mammals and endangered species. Each Action asks members to write personal letters to individuals in the corporation, government, or international organization responsible for the crisis or injustice. The public pressure exerted by a deluge of polite, informed letters has brought about policy changes on many occasions. It has helped stop oil drilling in Peru's Pacaya-Samiria National Reserve, preserve more than 2 million acres of rain forest on the Mosquito Coast of Honduras, and convince multinational companies to withdraw from Burma, a country with a record of abusing both human rights and the environment. Global Response also issues Young Environmentalist's Actions (YEAs), which teach geography, science, language, and letter writing, and can help young people develop the conscience and responsibilities of a global citizen. Both adult and youth members receive Action Status reports for updates on past Actions.

The Green Center
237 Hatchville Road
East Falmouth, MA 02536
(508) 564-6301
Web site: http://www.fuzzylu.com/greencenter/

The Green Center evolved from the legendary New Alchemy Institute, which for years researched environmentally sustainable agriculture, aquaculture, housing, landscaping, and energy generation. The Green Center distributes New Alchemy's publications and promotes those New Alchemy ideas with the most potential. It also administers New Alchemy's specialized library. The Green Center is in the process of setting up an educational nature center in the nearby town of Falmouth and a community-run organic farm.

Publications: Write for the publications list; it includes back issues of New Alchemy journals, technical bulletins, research reports, and working papers.

Greenpeace
1436 U Street NW
Washington, DC 20009
(202) 362-1177
Fax: (202) 462-4507
Web site: http://www.greenpeace.org/~usa

For 25 years, individuals working with Greenpeace have risked their lives in imaginative campaigns to defend the environment. The organization has become famous for the voyages of the *Rainbow Warrior* and the *MV Greenpeace.* Aboard those vessels, Greenpeace members have pursued whalers, then dived into the ocean to place themselves between endangered whales and the whalers' nets; they have anchored at ground zero to prevent undersea nuclear tests; and they have sailed to countries about to cut their old-growth forests and supported indigenous peoples' struggles against deforestation. Currently the organization focuses on stopping the factory-style overfishing that depletes marine ecosystems of their biodiversity and robs small-scale fishermen of their livelihood; deforestation of old-growth forests in tropical and temperate zones; proliferation of nuclear weapons and energy-generating stations; and environmental poisoning through production of toxic chemicals for which acceptable substitutes are available and burning of toxic waste.

Publications: A quarterly magazine is sent to members, and the organization's web site includes links to Greenpeace reports on marine biodiversity.

The Healing Forest Conservancy (HFC)
3521 S Street NW
Washington, DC 20007
Phone/Fax: (202) 333-3438
E-mail: moranhfc@aol.com

This organization was founded by Shaman Pharmaceuticals (see below) and works to protect the resources that Shaman depends on for its ethnobotanical research: biological diversity in tropical forests and the cultural and medicinal traditions of the indigenous peoples of the tropics. HFC accomplishes its mission by contributing money to local conservation efforts like the demarcation of native lands; the sustainable harvesting and marketing of natural forest products (which assigns live forests more economic value than the timber that could be sold if the forests were cut); the development of local markets for forest products; and the training of locals to collect, identify, and inventory specimens. Each year HFC presents the Richard Evans Schultes Award to a person or group of people who has made an outstanding contribution to the field of ethnobotany.

INBio
(Costa Rica's National Biodiversity Institute)
Instituto Nacional de Biodiversidad
Apdo. 22-3100 Santo Domingo de Heredia
Heredia
Costa Rica
011-506-244-2816
Fax: 011-506-244-0690
E-mail: askinbio@quercus.inbio.ac.cr
Web site: http://www.inbio.ac.cr

INBio has a four-part mission: to perform a species inventory on Costa Rica's protected areas, to promote the sustainable use of biodiversity by all sectors of Costa Rican society, to act as information managers in the cataloging of the inventoried species, and to disseminate knowledge of the biodiversity found in Costa Rica. Its bioprospecting program is one of INBio's most innovative activities. INBio has made agreements with the drug company Merck & Company, Inc. (see the Statistics, Illustrations, and

Documents chapter) and the Givaudan-Roure fragrance company to look for promising natural substances on their behalf. (Should these companies decide to develop a product based on any of the compounds INBio provides them, they will then draft a separate agreement with INBio regarding royalties.) This type of agreement breaks the traditional pattern in which poorer countries supply raw materials cheaply or for free to the industries of wealthier countries. The compensation paid to INBio will be shared with the country's system of protected areas to assure continued conservation in Costa Rica. For more detailed information on the agreement with Merck & Company, Inc., and INBio's mission, see the Statistics, Illustrations, and Documents chapter.

Institute for Local Self-Reliance (ILSR)
2425 18th Street NW
Washington, DC 20009-2096
(202) 232-4108
Fax: (202) 332-0463
E-mail: ilsr@igc.apc.org
Web site: http://www.ilsr.org

Founded in 1974, ILSR has worked to help local communities with economic development projects that utilize local resources and do not harm the environment. Initially ILSR worked with neighborhoods; later it added cities, regions, and entire nations. ILSR collaborates with grassroots organizations, local businesses, and local governments, providing them with the findings of the institute's research as well as technical assistance for projects and policy initiatives. Current projects include promoting the use of nonpolluting, plant-based fuels like ethanol, developing transportation alternatives that could wean Americans from their cars, showing that composting can become a major strategy for dealing with household and industrial organic wastes, and promoting the cultivation of industrial hemp in the Midwest.

Publications: ILSR staff members have published numerous papers and books on the institute's major issues. A publications list can be accessed through the ILSR web site.

International Alliance for Sustainable Agriculture (IASA)
1701 University Avenue SE
Minneapolis, MN 55414
(612) 331-1099
Fax: (612) 379-1527

E-mail: IASA@mtn.org
Web site: http://www.mtn.org/CAP

IASA is a 15-year-old organization that works to develop food production systems worldwide that are ecologically sound, economically viable, socially just, and humane. All of these conditions, in IASA's view, make up the concept of sustainability. The organization has worked at local, state, and national levels, and with the United Nations' Development and Environmental Programmes to develop plans and policies for sustainable agriculture. Through various networks it has established, it works with grassroots groups in Africa, Asia, and Latin America. IASA coordinates a campaign called Circle of Poison, which aims to stop the export from the United States of dangerous pesticides banned in this country but legal elsewhere.

Publications: MANNA is IASA's quarterly newsletter for members; several other publications are available, including *Breaking the Pesticide Habit*, by IASA founder Terry Gips.

Izaak Walton League
707 Conservation Lane
Gaithersburg, MD 20878
(301) 548-0150
E-mail: general@iwla.org
Web site: http://www.iwla.org/iwla/

This organization, whose primary focus is river and stream conservation, is made up of 50,000 individuals working at the grassroots level. It was formed in 1922 in Chicago when a group of 54 sportsmen became concerned about contamination of the country's top fishing streams and decided to take personal responsibility for their protection. They named their organization after a seventeenth-century angler-conservationist, and at the request of President Warren Harding organized the nation's first complete water pollution survey. For years the league has fought for clean water legislation. Its SOS (Save Our Streams) program helps grassroots activists monitor water quality in local streams and teaches about stream restoration and protection. Its wildlife protection work focuses on problems with freshwater fish. A new project works with farmers on sustainable agriculture practices and promotes a higher level of energy efficiency.

Publication: Outdoor Ethics is a free quarterly newsletter promoting safe and ethical outdoor recreation.

The National Audubon Society (NAS)
700 Broadway
New York, NY 10003
(212) 979-3000
Fax: (212) 979-3188
Web site: http://www.igc.apc.org/audubon/contents.html

More than a century old, the National Audubon Society was founded to protect birds at a time when great numbers were being killed for their plumes. Birds are still just as threatened as they were 100 years ago but by threats more difficult to combat: loss of habitat, use of pesticides, pollution of air and water, and shrinking wetland habitats. The National Audubon Society approaches its task in a variety of ways. It preserves habitat by managing a network of wildlife sanctuaries all over the United States, it helps develop conservation policies, it uses the court system to help enforce environmental laws, and it conducts scientific research. Current campaigns focus on links between overpopulation and loss of habitat, wetlands, restoration of the Everglades, endangered species, oceans, and forests. Audubon's extensive educational outreach includes camps for children and trips for adults.

Publications: Audubon is a bimonthly magazine for members; *Wetlands Forever* is the wetlands campaign's newsletter. NAS has produced more than two dozen videos on the environment, available through PBS Video (see Nonprint Resources chapter for address).

National Wildlife Federation (NWF)
1400 16th Street NW
Washington, DC 20036
(202) 797-6655 (legislative hotline)
Web site: http://www.nwf.org/nwf

The National Wildlife Federation is one of the oldest and largest conservation organizations in the United States. It engages in advocacy, education, and litigation for the conservation of wildlife, focusing on wetlands, water quality, endangered habitats, land stewardship, and sustainable communities. One of its longstanding programs is the Backyard Wildlife Habitat Program, which certifies privately owned wildlife refuges ranging in size from three window boxes on an office building to a 25,000-acre forest. Another innovative program is its Corporate Conservation Council, through which NWF staff meet with corporate executives from a variety of industries, opening doors between two sectors that have traditionally disagreed on many issues.

Publications: NWF publishes five magazines: *National Wildlife* and *International Wildlife*, covering conservation efforts and wildlife research worldwide; *EnviroAction*, an environmental news digest; *Ranger Rick* for elementary school–age kids; and *Your Big Backyard* for preschoolers. NWF also produces feature films on wildlife and conservation, publishes a conservation directory that lists hundreds of conservation organizations, and has developed a variety of environmental education curriculum materials.

Native Seeds/SEARCH
2509 N. Campbell Avenue, #325
Tucson, AZ 85719
(520) 327-9123
Fax: (520) 327-5821
Web site: http://desert.net/seeds/home.html

This organization "works to conserve the traditional crops, seeds, and farming methods that have sustained native peoples throughout the U.S. Southwest and northern Mexico." Its staff gathers, saves, and distributes seeds of this region's ancient crops and their wild relatives, does research on them, and educates the community about the plants' care and worth. Native Seeds considers that conservation of biodiversity is linked to the preservation of native cultures, and that "both are essential in our efforts to restore the earth." The organization offers free and discounted seeds to native gardeners and encourages them to maintain pure seed lines and establish their own personal seed banks. Besides offering seeds and training, Native Seeds runs a demonstration garden on the grounds of the Tucson Botanical Garden. It supports the traditions of "the original seed savers" (the Hopi, Zuni, Apaches, Tarahumara, and other peoples) by offering some of their crafts for sale and providing information about them in their publications.

Publications: An annual catalog offers seeds for such desert crops as teosinte, wild gourds, chiles, amaranth, and indigo. Members receive the quarterly newsletter, *The Seedhead News*, which contains features on native farmers and crops, gardening tips, book reviews, and information on the organization's workshops and events.

Natural Resources Defense Council (NRDC)
40 W. 20th Street
New York, NY 10011
(202) 727-2700

Fax: (202) 727-1773
E-mail: nrdcinfo@nrdc.org
Web site: http://www.nrdc.org/nrdc

This well-established environmental research and advocacy organization uses the legal system to assure that environmental protection laws are enforced. NRDC works in concert with local groups on many issues that relate directly to biodiversity, including acid rain, water pollution, pesticide policy, management of national forests and other public lands, conservation of endangered species, shoreline protection, habitat protection, desertification, deforestation, wetlands conservation, international wildlife trade, and international environmental treaties.

Publications: Members receive the quarterly *Amicus Journal* and the *NRDC Newsletter*, which comes out five times a year. Their web site has links to a variety of documents and reports prepared by NRDC staff.

The Nature Conservancy (TNC)
1815 N. Lynn Street
Arlington, VA 22209
(703) 841-5300
Fax: (703) 841-1283
Web site: http://www.tnc.org

The Nature Conservancy was founded in the 1950s and modeled after the British government's Nature Conservancy, whose mission was to buy land for nature preserves. Early TNC members plied wealthy friends for funds to buy and manage land that seemed "natural" enough to preserve. But in the 1970s, TNC adopted a biology-based approach, limiting its purchases to land that would help preserve genetic diversity—land that supported endangered species and ecological communities. Since then, TNC has become a world leader in conservation efforts, often working in partnership with governments. Its State Natural Heritage Program, established in 1974 with its first chapter in South Carolina, makes an inventory of species and classifies them according to degree of endangerment. Now a State Natural Heritage Program operates in every U.S. state, and several Conservation Data Centers operate in Latin America. This methodical data collection system helps TNC and other entities prioritize their conservation efforts.

Publications: Nature Conservancy is a bimonthly magazine for members. The international division of TNC publishes *Beyond Our Borders* quarterly, and each state chapter publishes its own newsletter.

The New York Botanical Garden (NYBG)
Bronx, NY 10458-5126
(718) 817-8700
Fax: (718) 220-6504
Web site: http://www.nybg.org

The New York Botanical Garden has recently found itself at the forefront of the current rush to explore and document botanical biodiversity. But for the past century, long before this pursuit became popular, scientists and graduate students from NYBG have been combing the wilds of the world and returning with plant specimens and information about how native peoples use them. Because of its headstart in the search, the NYBG now has invaluable ethnobotanical stores and priceless data. NYBG is divided into divisions that tackle different tasks associated with biodiversity. The Institute of Systematic Botany inventories plants, organizes data, and disseminates it. Botanists affiliated with the Institute of Economic Botany study the relationship between plants and people in 18 tropical countries. The field of economic botany is especially important in conservation, because if a sustainable use of wild plants can be achieved, local peoples will have a great incentive to conserve wild habitat.

Publications: NYBG publishes numerous technical journals, monographs, and books. Write to the Scientific Publications Department c/o NYBG for its catalog.

The Orion Society
195 Main Street
Great Barrington, MA 01230
(413) 528-4422
Fax: (413) 528-0676
E-mail: orion@bcn.net

The Orion Society uses the work of writers, naturalists, and educators to help people recognize and better understand their own connection to nature. It hopes that the rediscovery "of our human bond with the natural world and with each other," especially our intimate connection to our own place, will lead to the renewed health of local ecosystems, towns, and cities. The society works with teachers to develop vanguard environmental education curricula and sends nature writers and poets on tour across the country. It also develops conferences, including on-line conferences, designed to connect grassroots organizations.

Publications: Orion Society Notebook is a quarterly that showcases the efforts and visions of educators, writers, and activists. *Orion* magazine publishes essays, fiction, poetry, and book reviews that contemplate the relationship between people and nature. Write for a catalog of available back issues, each with a specific theme (land use, nature and healing, hunting, rivers, national parks, etc.).

The Population Institute (PI)
107 2nd Street NE
Washington, DC 20002
(202) 544-3300
Fax: (202) 544-0068
E-mail: popline@primanet.com
Web site: http://www.populationinstitute.org/

The links between overpopulation and global environmental problems like deforestation, water scarcity, and famine are becoming more obvious every day. The Population Institute, founded in 1969, works toward stabilization of global population growth rates, especially in regions where overpopulation is most serious. Its strategy includes helping empower women—especially adolescents—so they can take control of their own fertility, and providing family planning information and products to women of childbearing age. PI lobbies the U.S. Congress to support population control efforts in countries with steep growth rates. Although population control is a priority for most of the world and the United Nations, the United States is last among the 20 richest nations to support population control efforts within its borders and beyond.

Publication: POPLINE is a bimonthly newsletter and news and feature service serving more than 2,100 daily newspapers around the world. It provides analyses of worldwide population control policies and gives up-to-date facts on overpopulation.

Rachel Carson Council, Inc.
8940 Jones Mill Road
Chevy Chase, MD 20815
(301) 652-1877
Fax: (301) 951-7179
E-mail: rccouncil@aol.com
Web site: http://members.aol.com/rccouncil/ourpage/rcc_page.html

This organization was founded in memory of the author of *Silent Spring*, a wake-up call to 1950s U.S. society about the effects of widely used pesticides (see Print Resources and Biographical Sketches chapters). The council is dedicated to "fostering a sense of wonder and respect toward nature and helping society realize

Rachel Carson's vision of a healthy and diverse environment." In practice, it focuses on chemical contamination and serves as a clearinghouse for information on pesticides for scientists and the general public. Anyone with questions on pesticides (for example, dangers associated with a specific product) can request information from the council.

Publications: Basic Guide to Pesticides (1992) is a summary of the council's many years of research on the topic. More than 40 more publications can be ordered from the council. Members receive a periodic newsletter.

Rainforest Action Network (RAN)
221 Pine Street, Suite 500
San Francisco, CA 94104
(415) 398-4404
Fax: (415) 398-2732
E-mail: rainforest@ran.org
Web site: http://www.ran.org

RAN's mission is "to preserve rain forests and to protect the rights of indigenous people living in and around them." Of the many organizations working to stop rain forest destruction, RAN can be distinguished for its focus on grassroots efforts and citizen action. It supports the efforts of local activists in 60 countries worldwide, working especially with indigenous and environmental groups in tropical countries. Its Protect an Acre program donates money to indigenous forest dwellers who use it to obtain legal title to their land and demarcate its borders. RAN also specializes in coordinating actions such as boycotts; there are 150 Rainforest Action Groups (RAGs) throughout the country that can mobilize thousands of local citizens for emergency campaigns. Its first campaign in 1987 was a successful boycott of Burger King, which had been importing millions of dollars in beef from tropical countries where forests had been cut down to make room for cattle pastures. Currently RAN is asking people to boycott Mitsubishi for its indiscriminate logging practices and to boycott tropical timber and timber products in general unless they come from verified sustainable logging operations. To publicize the plight of rain forests, RAN sponsors conferences and conducts frequent media campaigns, which sometimes include full-page ads in the *New York Times*.

Publications: A monthly *Action Alert* and a quarterly *World Rainforest Report* inform members of current rain forest status and actions that can be taken.

The Rainforest Alliance (RA)
65 Bleeker Street
New York, NY 10012
(212) 677-1900
Fax: (212) 677-2187
E-mail: canopy@ra.org
Web site: http://www.rainforest-alliance.org

Tropical rain forests benefit everyone on earth and are everyone's responsibility, according to the Rainforest Alliance credo. Unfortunately, people living in countries that harbor the world's tropical rain forests are among the world's poorest and most desperate for a decent livelihood. RA develops and promotes economic alternatives for human rain forest inhabitants and their neighbors—alternatives that are environmentally sound, economically viable, and socially desirable. Two programs encourage a partnership between industries and environmentalists. The SmartWood program, in existence since 1989, certifies lumber projects that manage their forests well, have good labor and community relations, protect the biodiversity of their area, and reforest. More than 7 million acres have been certified so far. The ECO-OK programs develop ways to grow crops while doing the least possible damage to local waterways and ecosystems. Bananas are currently being certified, and RA is developing guidelines to certify oranges, coffee, cocoa, Brazil nuts, and vanilla. Other activities include a Natural Resources and Rights Program, which helps indigenous peoples gain a greater voice in the formation of environmental policies; an Amazon Rivers program, where scientists work with the governments of Amazon Basin countries to protect that huge and vitally important habitat; and a Conservation Media Center, based in Costa Rica, which issues newsletters and briefs to conservationists in Latin America.

Publications: The Canopy is RA's membership newsletter. The Conservation Media Center publishes the *Eco-Exchange* newsletter in English and its Spanish-language version, *Ambien-tema.*

The Resource Foundation
P.O. Box 3006
Larchmont, NY 10538-9908
Phone/Fax: (914) 834-5810

The Resource Foundation acts as a partner for grassroots self-help organizations in Latin America, many of which work to establish a more harmonious relationship between people and the environment. The foundation seeks large, tax-deductible donations, which it channels to its Latin American partners. It also provides them with technical assistance when needed. The Resource Foundation funds projects that benefit large numbers of poor people while keeping expenditures relatively low. Some recent beneficiaries include the Agro-Ecological Technical School in Pirque, Chile, which trains its students in sustainable agriculture; a program in Bolivia to help small farmers restore the fertility of their soils through erosion control and reforestation; and the Family Tree Project in Nicaragua, which educates families about deforestation and how to cultivate trees, and distributes fruit trees for families to plant and tend. By helping improve the lives of poor people in developing nations, the organization helps relieve pressure on natural resources.

Publication: A biannual newsletter, *Resource Foundation News*, updates donors on project accomplishments.

Resource Renewal Institute (RRI)
Fort Mason Center, Building A
San Francisco, CA 94123
(415) 928-3774
Fax: (415) 928-6529
E-mail: info@rri.org
Web site: http://www.rri.org

RRI's expertise lies in helping countries, states, regions, and cities design Green Plans: comprehensive, long-term environmental strategies leading to true sustainability. Green Plans are led by governments in partnership with industry and with sensitivity toward public concerns. In order for them to be successful, they must have clear goals and timetables as well as an adequate budget. All countries signing the Agenda 21 that came out of the 1992 Rio Earth Summit agreed to implement Green Plans as a step toward sustainable development. So far, however, only three countries (New Zealand, the Netherlands, and Singapore) and several cities have adopted them. In those places a consensus has been achieved that encourages all participants to cooperate, and bureaucracy has been reduced because all laws pertaining to the environment have been consolidated. In addition to its work in green plans, RRI reaches out to corporations, state governments, and the general public in the United States; it leads Seeing Is Believing tours to Green Plan communities. It was recognized with the 1996 President's Sustainable Development Award.

Publications: RRI staff members have written two books describing their work: *Saving Cities, Saving Money* (by John Hart, AgAccess, 1992) and *Green Plans: Greenprint for Sustainability* (by Huey Johnson; see Print Resources chapter). A video entitled *Green Plans* (see Nonprint Resources chapter) also explains the concept in detail.

Rodale Institute
611 Siegfriedale Road
Kutztown, PA 19530-9749
(610) 683-1400
Fax: (610) 683-8548
Web site: http://www.fadr.msu.ru/rodale/

The Rodale Institute was founded 50 years ago by J. I. Rodale, who researched and promoted the relation between soil health and disease prevention. The younger Rodales now base their work on a three-way connection: healthy soil = healthy food = healthy people. This concept ties into biodiversity conservation because if farmers can ensure continuously high productivity of their land, they will not follow the environmentally destructive pattern common throughout the world: inappropriate farming techniques ruin land; ruined land forces farmers to move into pristine areas and convert natural habitat into more farmland. The institute's activities are too numerous to list in entirety, but here are a few: In the United States, Rodale works with farmers and backyard gardeners to reduce their dependence upon chemical fertilizers and pesticides. Abroad, they help farmers reach maximum productivity with minimum environmental destruction. A Senegal project helps women farmers produce without pesticides. In the Petén region of Guatemala, Rodale's Centro Maya project is reversing destruction of rain forest by teaching colonists the ancient Mayan farming methods that renew fertility of soil and naturally prevent gardens from being overrun by jungle vines. And in Russia, Rodale Institute technicians are easing the transition from large state-run cooperative farms to small-scale family-run farms. The institute farm in Kutztown, Pennsylvania, is a model farm open to visitors.

Publications: The institute publishes numerous technical papers and journals for farmers, agronomists, and development workers. These are listed in their catalog, available upon request. Members receive the *Partner Report* newsletter, published three times a year.

Seed Savers Exchange
3076 North Winn Road
Decorah, IA 52101
(319) 382-5990

This is an organization of serious backyard gardeners concerned with the eroding genetic and species diversity of seeds available from traditional seed catalogs. They are concerned about preserving "heirlooms," unusual plants whose seeds were traditionally saved from year to year and passed down through the generations. Members grow their own families' heirloom varieties, traditional Native American crops, varieties developed and grown by Mennonite and Amish gardeners, little-known foreign varieties, and vegetables that most commercial seed companies have dropped because they are too unusual. Members save their own seeds after every growing season and offer them to other gardeners free, through an annual yearbook. The headquarters is located on the grounds of Heritage Farm, where thousands of vegetable varieties are grown, the most diverse orchard in the country produces about 700 types of apples, and a rare breed of wild cattle roams pastures. Members meet at the farm every July for a weekend convention with talks, slide shows, demonstrations, workshops, and garden-fresh meals.

Publications: Seed Savers issues three annual publications. January's *Seed Savers Yearbook* is the catalog through which members offer their heirloom seeds to other gardeners. The *Summer Edition* contains reports of projects at Heritage Farm, profiles of some of the best heirlooms, genetic preservation information, etc. The *Harvest Edition* includes manuscripts from the past summer's convention talks and a plant finder service for locating lost varieties.

Seeds of Change
P.O. Box 15700
Santa Fe, NM 87506-5700
(505) 438-8080
Fax: (505) 438-7052
E-mail: gardener@seedsofchange.com
http://www.seedsofchange.com

Organic seed supplier Seeds of Change delights gardeners with its multitude of rare varieties and at the same time serves an important role in biodiversity conservation. It sells seeds for grains, legumes, fruits, and vegetables that are little known but present distinct agricultural, environmental, and nutritional advantages. For example, amaranth is a nutty grain that grows abundantly on huge seedheads. Though it was a staple and had religious significance for indigenous Americans, it was all but decimated by the Spaniards. The tepary bean is a nutritional but undercultivated bean from the American Southwest. The mission of Seeds of Change is to continue rescuing "lost" crops, to allow natural evolution to continue improving them, and to promote their reintegration into our diet.

Publication: An annual seed catalog describes the varieties sold by Seeds of Change.

Shaman Pharmaceuticals, Inc.
213 East Grand Avenue
South San Francisco, CA 94080-4812
(415) 952-7070
E-mail: jcossmon@shaman.com

This innovative company searches the tropics for new drugs through the methods of ethnobotany—the study of plant use by different cultures. By analyzing the medicinal plants used by native peoples of the tropics, Shaman's work is more efficient and cost-effective than that of drug companies that screen plants indiscriminately. Shaman reported in 1996 that of the 800 plants it had screened, more than half showed promise as potent therapeutic drugs, including one that may heal outbreaks of genital herpes. Shaman is also unique in the corporate pharmaceutical world in the reciprocity programs it has initiated. Communities that assist Shaman are given up to 15 percent of the money budgeted for the expedition to help with specific grassroots projects, such as demarcation of traditional lands, building or remodeling traditional medical centers in communities, installing potable water systems, and supporting community health projects. Long-term help is offered in other ways: bringing local scientists to Shaman's labs to learn technical skills, providing lab equipment and materials, and funding ethnobotanical studies and local research. Shaman founded a sister nonprofit, the Healing Forest Conservancy (see separate listing above), to provide even longer-term compensation.

Sierra Club
85 Second Street
Second Floor
San Francisco, CA 94105
(415) 977-5653
E-mail: information@sierraclub.org
Web site: http://www.sierraclub.org

One of America's original environmental organizations, the Sierra Club was founded in 1892, electing legendary naturalist John Muir its first president. Its statement of purpose reads, "To explore, enjoy and protect the wild places of the Earth; to practice and promote the responsible use of the Earth's ecosystems and resources; to educate and enlist humanity to protect and restore the quality of the natural and human environment; and to use all lawful means to carry out these objectives." The club's half a million members form more than 60 chapters and almost 400 local groups working at both local grassroots and national levels. The Environmental Protection Agency (EPA), the Endangered Species Act, the Arctic National Wildlife Refuge, and the Clean Air and Clean Water Acts were all established thanks to a strong Sierra Club lobby in Congress; many national parks, such as Grand Canyon, Yosemite, and Mount Rainier, were founded or expanded with a push from the Sierra Club.

Publications: Sierra is a magazine available to members; *The Planet* is a resource for activists.

Society for Conservation Biology
Department of Biological Sciences
Stanford University
Stanford, CA 94305
E-mail: conbio@igc.org

Conservation biology is a relatively new academic field defined two decades ago by biologist Michael Soulé (see Biographical Sketches). It recognizes the crucial ties between biology and conservation and, in turn, their relation to other disciplines that influence conservation, such as economics, philosophy, law, and the social sciences. The Society for Conservation Biology is an academic organization that seeks to "to help develop the scientific and technical means for the protection, maintenance, and restoration of life on the planet—its species, its ecological and evolutionary processes, and its particular and total environment." It promotes high-quality research, publishes and disseminates information, encourages communication between all related fields, funds studies, and recognizes outstanding contributions.

Publication: Members receive the quarterly *Conservation Biology*, which includes articles on conservation from a variety of perspectives.

Society for Ecological Restoration (SER)
1207 Seminole Highway
Madison, WI 53711
(608) 262-9547
Fax: (608) 265-8557
E-mail: ser@vms2.macc.wisc.edu
Web site: http://nabalu.flas.ufl.edu/ser/SERhome.html

The Society for Ecological Restoration unites professional ecologists and restorationists with committed lay practitioners of ecosystem restoration. SER members are engaged in many restoration projects in areas as varied as Philadelphia's streams and brooks, an abandoned mine in Utah, a bulldozed section of the Indiana Dunes, and a former army base in Colorado. Restoring despoiled places is important for scientists, who usually observe ecosystems as outsiders but with these projects can experiment in their re-creation. Restoration also offers great benefits for laypeople. Nature has always been a healing force for humans, but a restoration project offers a way for people to perform an integral role in nature's own healing. Despite its strong advocacy of restoration, SER firmly maintains that restoration cannot be considered an alternative to conservation of existing ecosystems.

Publications: SER publishes two journals. *Restoration Ecology* is a quarterly scientific journal of restoration theory and research. *Restoration & Management Notes* is a semiannual guide for restorationists. The quarterly newsletter, *SER News*, focuses on SER activities.

Survival International
11-15 Emerald Street
London WC1N 3QL
England
0171-242-1441
Fax: 0171-242-1771
Web site: http://www.survival.org.uk/

Founded in 1969, Survival International works with indigenous peoples' organizations all over the world but prefers to focus on the most vulnerable: those who have just come into contact with outside civilization. Problems for indigenous peoples come in a variety of forms: encroachment by outside settlers, racism, mining, assassinations, oil exploration, road building, disease, military invasions, poverty, and misguided conservation efforts. Survival responds to urgent threats with letter-writing campaigns to exert public pressure on governments, companies, financial institutions, extremist missionaries, and guerrilla groups. Its educational mission involves modernizing the traditional Western view of indigenous people as primitive relics. Survival also facilitates networking among indigenous groups so that they may learn from one another.

Union of Concerned Scientists (UCS)
2 Brattle Square
Cambridge, MA 02238-9105
(617) 547-5552
E-mail: ucs@ucsusa.org
Web site: http://www.ucsusa.org/

UCS was born at the Massachusetts Institute of Technology in 1969, out of a movement of faculty and students concerned about what they felt was a misuse of science and technology. Those fields, they argued, should be used to help solve environmental and social problems rather than to strengthen military capabilities. Over the years, UCS has focused on issues of urgency: nuclear proliferation, global warming, and the degradation of biodiversity. Through their Sound Science Initiative, a network of scientists presents credible and timely scientific information about topics such as biodiversity, climate change, ozone depletion, and population growth to policy makers, the media, and the general public. The 1992 Union of Concerned Scientists' World Warning to Humanity (see the Statistics, Illustrations, and Documents chapter), signed by 1,700 scientists, including 100 Nobel laureates, describes worsening threats to the natural resources that sustain life on earth.

Publications: UCS publications include a quarterly magazine, *Nucleus*, and many other books, reports, papers, brochures, and videos.

United Nations Environment Programme (UNEP)
P.O. Box 30552
Nairobi, Kenya
254-2-62-1234/3292
Fax: 254-2-62-3927/3692
E-mail: ipainfo@unep.org
Web site: http://www.unep.org

UNEP was born out of the 1972 Stockholm Conference on the Human Environment and was established to provide an "environmental conscience" for the United Nations (UN). A huge coordinating organization that oversees projects undertaken by other UN agencies, it has always integrated economic development with environmental protection. Some of its many areas of interest related directly to biodiversity include marine pollution, deforestation, wildlife conservation, sustainable agriculture, climate change, desertification, "green" technologies, ecotourism, and natural resource accounting (calculating environmental costs into product and service prices).

Publications: UNEP collaborates with the World Bank and the World Resources Institute (see below) to produce the biannual *World Resources Report*. Other major publications are the *State of the Environment* reports. It produces many other publications as well, listed in each of its annual reports.

United States Fish and Wildlife Service (USFWS)
Department of the Interior
1849 C Street NW
Washington, DC 20240
(202) 208-5634
Web site: http://www.fws.gov/

The mandate of the USFWS is to conserve, protect, and enhance the habitat of fish and wildlife in the United States. Its specific assignments include the protection of migratory birds and endangered species, certain marine mammals, and freshwater and anadromous fish (fish like salmon that spend most of their life in the ocean but breed in fresh water). It is the agency charged with implementing the Endangered Species Act of 1973. The USFWS identifies species that are "endangered" (likely to disappear) or "threatened" (likely to become endangered), places them on the lists, and develops and oversees recovery plans for them. Currently about 700 species are listed. Some species, such as the American alligator, the bald eagle, and the California gray whale, have enjoyed remarkable recoveries. USFWS is also the agency in charge of upholding international wildlife and wildlands accords and treaties, such as the Ramsar wetlands protection agreement, the Convention on International Trade in Endangered Species (CITES), and migratory bird treaties. The USFWS is in charge of wetland protection and manages an extensive system of National

Wildlife Refuges. In addition to these publicly owned refuges, USFWS works with 11,000 landowners in its Partners for Wildlife program to restore habitat on private property.

Publications: USFWS maintains 40 Internet servers that provide complete information about the agency's many tasks.

The Wilderness Society
900 17th Street NW
Washington, DC 20006
(202) 833-2300
Fax: (202) 429-2658
Web site: http://www.lbbs.org/wild/about.htm

The Wilderness Society was founded to promote widespread adoption of the land ethic Aldo Leopold articulated in his book, *A Sand County Almanac* (see Print Resources chapter). Leopold said a land ethic "changes the role of *Homo sapiens* from conqueror of the land-community to plain member and citizen of it. It implies respect for his fellow-members and also respect for the community as such." The Wilderness Society's work has centered on legislation; it has led the pack on most of the major public lands issues since its founding in 1935, and it has helped legislate many protection acts as well.

Publications: Wilderness is a magazine published quarterly. Several other newsletters focus on endangered species, forests of the northwestern United States, and the Arctic Wildlife Refuge. Numerous reports are available on a variety of topics. Its web site provides a guide to other environment-oriented web sites, with many links and lots of beginner-friendly information.

Wildlife Preservation Trust International (WPTI)
3400 West Girard Avenue
Philadelphia, PA 19104-1196
(215) 222-3636
Fax: (215) 222-2191
E-mail: WPTI@aol.com
Web site: http://www.columbia.edu/cu/cerc/wpti.html

WPTI focuses on saving endangered species throughout the world, especially in tropical countries because of their tremendous biodiversity. Four hundred conservationists, called "New Noahs" because they tend modern-day conservation "arks," have been trained by WPTI. They work in the trust's 27 projects in 15 countries, including Cuba, Belize, Indonesia, and Madagascar. Among the species targeted for study, captive breeding, health care, reintroduction, and transfer to safer habitats are tapirs, parrots, golden lion tamarins, orangutans, and lemurs. Most projects include an educational component, because WPTI, like many other conservation organizations, believes that conservation efforts are futile unless local residents see the benefits of coexisting with nature. Their Internet-based The Wild Ones club (http://www.columbia.edu/cu/cerc/WildOnes) is an interactive program that allows children from 20 countries to learn about nature and communicate with one another.

Publication: On the Edge is WPTI's newsletter.

World Conservation Monitoring Centre (WCMC)
219 Huntingdon Road
Cambridge CB3 ODL
United Kingdom
44 (0)223 277314
Fax: 44 (0)223 277136
E-mail: info@wcmc.org.uk
Web site: http://www.wcmc.org.uk/

Founded in the early 1980s by the three world conservation giants—IUCN, UNEP, and WWF—the World Conservation Monitoring Centre serves as a source of information and technical support on global biodiversity conservation and sustainable development issues. Its current projects include helping countries make their own biodiversity assessments, establishing national biodiversity centers, identifying countries with high numbers of endemic species, and monitoring indicators of ecosystem quality and conditions.

Publications: Valuable country-by-country biodiversity information is available through WCMC's homepage on the World Wide Web. A list of publications is also provided on the web site.

The World Conservation Union (IUCN)
Rue Mauverney 28
CH-1196 Gland, Switzerland
41-22-999-00-01
Fax: 41-22-999-00-02
Web site: http://www.iucn.org/

The IUCN, founded in 1948, is a partnership of more than 800 states and governmental and nongovernmental organizations from 125 countries. Its primary mission is to promote conservation in concert with economic development in the quest for a sustainable standard of living for the world's entire population. IUCN's more than 6,000 expert consultants have assisted many countries to prepare national conservation strategies and provided them with scientific knowledge, technical support services, and strategy ideas for conservation projects. Its central headquarters in Switzerland represents its members on the world stage and coordinates the worldwide activities of the IUCN's six commissions: Ecology, Education and Communications, Environmental Law, Environmental Strategy and Planning, National Parks and Protected Areas, and Species Survival.

Publications: IUCN publishes many books and journals on topics such as biodiversity and sustainable development, threatened species, protected lands, forests, wetlands, marine and coastal areas, and environmental law. Their well-known *Red Data Books* list and describe species considered threatened, to varying degrees, with extinction. These books are published for specific taxa (such as birds) or for specific countries or regions of the world (such as southern Africa).

World Resources Institute (WRI)
1709 New York Avenue NW
Washington, DC 20006
(202) 638-6300
Fax: (202) 638-0036
Web site: http://www.wri.org/wri/

WRI recognizes that the well-being of human society depends upon the health of its environment and works to help human society use the earth's natural resources at a rate that will assure their preservation for future generations. In its role as a think tank, WRI analyzes scientific findings, economic situations, and practical experience. It provides information based on this research to governments and private organizations in more than 50 countries that are grappling with environmental and development challenges. WRI works with poor countries where the main problems are deteriorating natural resources and overpopulation. One of its priorities is the conservation of biodiversity. WRI's Biological Resources and Institutions program has three main strategies: (1) integrating the conservation of biological resources into national development plans, (2) developing new conservation strategies in which the cost of conservation is shared by all parties that benefit from conservation, and (3) developing creative ways to halt the destruction of biodiversity.

Publications: An annual reference book, *World Resources*, gives information on every country's resources. WRI's web site has links to many of the Institute's articles on biodiversity.

World Wildlife Fund (WWF)
1250 24th Street NW
Washington, DC 20037
(202) 293-4800
Web site: http://www.panda.org/home.htm

World Wildlife Fund is the American branch of the international WWF network, which extends throughout 50 countries on five continents. WWF works in nearly 100 countries to preserve biodiversity and promote sustainable development. Since it was founded in 1961, it has helped establish and manage more than 450 national parks and reserves worldwide. In recent years, WWF has realized that conservation will not succeed without local support, so it has brought environmentally friendly economic development programs to villages, neighboring parks and reserves and has trained locals as conservation specialists so that they can take more responsibility in managing their local wild area. Because of its international coverage, WWF learns right away about urgent threats to endangered species and can act quickly to mobilize massive campaigns. WWF also works on indirect threats to global biodiversity, such as climate change and abuse of toxic chemicals, and collaborates with many other international organizations, including the IUCN and the UNEP.

Publication: A catalog of WWF's extensive scientific and educational publications is linked to the WWF homepage on the Internet.

Worldwatch Institute
1776 Massachusetts Avenue NW
Washington, DC 20036-1904
(202) 452-1999
Fax: (202) 296-7365
E-mail: worldwatch@igc.apc.org

Worldwatch Institute, dedicated to fostering a society that meets human needs in a manner that does not compromise the health of the environment or future generations, evaluates information and generates policy advice on environmental issues such as climate change, energy efficiency, population growth, food production, and water resources. Its numerous publications are widely available and very readable, which allows them to be of use, according to an institute statement, to "a couple trying to decide whether to have another child or to a cabinet minister shaping a national energy strategy."

Publications: The institute's publications are translated into several languages and distributed to policymakers throughout the world. The best known is the annual *State of the World* report. Each edition identifies ten new environmental trends and assesses progress toward the sustainable society the institute envisions. Other publications include *Vital Signs, World Watch* magazine, the Worldwatch Papers, the Environmental Alert series, research briefs, and the Worldwatch database disk.

The Xerces Society
4828 Southeast Hawthorne Boulevard
Portland, OR 97215
(503) 232-6639
Fax: (503) 233-6794

Invertebrates such as insects, spiders, centipedes, and mites make up between 90 and 95 percent of all animal species and perform invaluable services for all species on the planet. They cycle nutrients, pollinate most types of flowering plants, disperse seeds, maintain the structure and fertility of the soil, keep populations of other organisms under control, and serve as a source of food for others. The Xerces Society formed to educate people about these important organisms and to work specifically toward their conservation by protecting their habitat. It has also pioneered a new technique for monitoring the health of aquatic habitats: documenting the presence or absence of insects in their larval forms—food for many fish. The society took the name of the Xerces Blue butterfly, the first butterfly in North America known to have become extinct due to human impact.

Publications: Xerces publishes a biannual magazine called *Wings: Essays on Invertebrate Conservation*. It has also authored a book entitled *Butterfly Gardening: Creating Summer Magic in Your Garden* (Sierra Books, 1990).

Zero Population Growth (ZPG)
1400 16th Street NW, Suite 320
Washington, DC 20036
(202) 332-2200
Fax: (202) 332-2302
Web site: http://www.zpg.org

Zero Population Growth combats human-generated impact on nature by advocating the stabilization of population growth in the United States and abroad. Its activities include publicity campaigns, curricula development, lobbying, and helping develop pertinent public policy.

Publications: ZPG Reporter is a bimonthly newsletter on population politics and research and ZPG activities. *The Activist* is a quarterly publication suggesting actions for local activists. Curriculum materials are available for teachers.

Print Resources

The past decade has seen a surge in print publications addressing the multiple facets of biodiversity. This chapter describes a fraction of those books, focusing on the most recent and on those most accessible to a general readership.

In addition to the vast number of books addressing biodiversity, many periodicals frequently contain articles pertinent to biodiversity. By searching the databases listed in the Nonprint Resources chapter, you will discover journals and magazines that frequently address your topic of interest. *Science Times,* which is included in the *New York Times* every Tuesday, almost always includes articles pertinent to the scientific aspects of biodiversity.

The books listed in this chapter are divided into 12 subject areas:

- **Atlases, Overviews, and Reports** includes books that are comprehensive in scope and often include the work of many contributors.
- **Anthologies and Conference Proceedings** includes some of the more technical information; the chapters in these books usually address very specific topics.

- **Ethics and Biodiversity** ponders biodiversity from a humanist perspective: our role in its destruction and its preservation.

- **Biodiversity Detailed: Focus on Specific Ecological Communities, Taxonomic Groups, or Scientific Questions** describes biodiversity from a scientific viewpoint and includes ecosystem descriptions, portraits of certain classes of organisms, and discussions of certain scientific questions pertaining to the new field of biodiversity.

- The books listed in **Biodiversity in Crisis,** while containing information about biodiversity itself, focus on habitat destruction and the ensuing extinctions, both in ancient history and at the present time.

- **Indigenous Peoples and Biodiversity** spotlights the sustainable lifestyle led by traditional indigenous peoples in natural ecosystems throughout the world. Living in this way, indigenous peoples depend directly upon biodiversity at the same time that they actively conserve it.

- **Agriculture and Biodiversity** addresses the issue of genetic erosion in the world's main crops and gives some proactive responses to it.

- **Legal and Legislative Responses to the Biodiversity Crisis** discusses international treaties and strategies for making conservation legislation as effective as possible.

- ***In-situ* Conservation Approaches** describes wildlands management techniques and gives proposals and examples from different countries.

- ***Ex-situ* Conservation Approaches** features captive breeding programs and botanical gardens.

- **Restoration of Ecosystems** discusses attempts at re-creating lost or damaged ecosystems.

- **Grassroots Actions in Response to the Biodiversity Crisis** provides ideas for how nonspecialists can contribute to recovery from the serious decline in biodiversity.

Atlases, Overviews, and Reports

Baskin, Yvonne. **The Work of Nature: How the Diversity of Life Sustains Us.** Washington, DC: Island Press, 1997. 263 pages. ISBN 1-55963-519-3

Commissioned by the Scientific Committee on Problems of the Environment (SCOPE), a project headed by some of the world's top biodiversity specialists, this book explores biodiversity not from a species-by-species or potentially useful product-by-product perspective but rather from the standpoint of the services provided by intact ecosystems. Rain forests, deserts, tundras, coral reefs, and other ecosystems provide the world with clean water, fertile soil, pollination, generation of game, etc. By translating the documents generated by SCOPE researchers into language that a general reader can absorb, this book details the probable consequences of a widespread degradation of biodiversity.

Braus, Judy, editor. **WOW! A Biodiversity Primer.** Washington, DC: World Wildlife Fund, 1994. 68 pages. ISBN 0-89164-142-4

This magazine-format introduction for elementary and middle school students shows that biodiversity is everywhere and that we depend upon it in more ways than we realize. Its varied contents include an interview with "Dr. B" (E. O. Wilson); a "Natural Inquirer" segment featuring fantastic phenomena such as exotic male dancing birds of paradise in Papua New Guinea and a killer with no brain or heart (the Portuguese man-of-war); profiles of people who have fashioned biodiversity conservation into a career; how much biodiversity is found in the products sold at a typical shopping mall; short stories; the musicians who make up Seattle's Ecosound environmental band; and much more. *WOW!* and its educator's guide can be purchased in bulk from WWF (see contact information in the Organizations chapter).

Durrell, Lee. **State of the Ark: An Atlas of Conservation in Action.** New York: Doubleday, 1986. 224 pages. ISBN 0-385-23668-9

This encyclopedic oversize volume covers a panoply of biodiversity-related matters: description of ecosystems and the web that connects them internally and to one another; assessments of humanity's myriad effects on nature; a region-by-region list of endangered habitats, human pressures, and development issues; and the story of how a worldwide conservation movement has risen to heal an ailing planet.

Groombridge, Brian, editor. **Global Biodiversity: Status of the Earth's Living Resources.** London: Chapman and Hall, 1992. 594 pages. ISBN 0-412-47240-6

Compiled by the World Conservation Monitoring Centre (see the Organizations chapter), this sourcebook describes biodiversity in scientific terms, lists its uses and values, and describes conservation and management practices. It provides hundreds of charts and maps on a wide variety of topics and gives extensive coverage to global extinctions.

Heywood, V. H., editor, and R. T. Watson, chair. **Global Biodiversity Assessment.** Cambridge, UK: Cambridge University Press, 1995. 1,140 pages. ISBN 0-521-564808

An authoritative encyclopedia of biodiversity, this comprehensive volume is two and a half inches thick, with hundreds of contributors writing about thousands of subtopics of biodiversity. If read in full, the book will provide a complete education, from scientific fundamentals such as taxonomy, genetics, and ecosystem function, to the distribution of taxa throughout the world, to the different ways of valuing biodiversity, to inventory projects, to conservation methods.

Lovelock, James. **Healing Gaia: Practical Medicine for the Planet.** New York: Harmony Books, 1991. 192 pages. ISBN 0-517-57848-4

Based on the metaphor of Gaia (planet earth) as an ailing patient, this book takes the planet's pulse, checks her vital signs, and recommends treatment. The chapters examine Gaia from a variety of perspectives (anatomy, physiology, biochemistry, climate regulation, etc.), consistent with Lovelock's integrative approach to science. (See the author's biography in the Biographical Sketches chapter.)

McNeely, Jeffrey A., Kenton R. Miller, Walter V. Reid, Russell A. Mittermeier, and Timothy B. Werner. **Conserving the World's Biological Diversity.** Gland, Switzerland: IUCN; and Washington, DC: WRI, CI, WWF-US, and the World Bank, 1990. 193 pages. ISBN 0-915825-42-2

Five of the world's most powerful actors in conservation—the World Conservation Union (IUCN), the World Resources Institute (WRI), Conservation International (CI), the World Wildlife Fund-US (WWF-US), and the World Bank—collaborated on this volume, which is peppered with charts, lists, graphs, and reprinted documents. The first chapters describe biodiversity, its value, and the threats to it. Treated more extensively in the next chapters are methods for conserving it. The first half of the book gives general information, and the second is designed for conservationists: it shares experiences, analyses, and practical approaches for governments as well as nongovernmental information.

Myers, Norman, editor. **Gaia: An Atlas of Planet Management.** Rev. ed. New York: Doubleday, 1993. 272 pages. ISBN 0-385-42626-7

A comprehensive portrait of conservation issues on planet earth, this atlas is divided into chapters addressing land, ocean, elements, evolution, humankind, civilization, and management. Each chapter is introduced by a renowned leader in its subject area and includes social, political, and scientific information. Diagrams, maps, and charts detail concerns such as clean water supply, nuclear buildup, the hole in the ozone layer, fishery depletion, and the causes and effects of poverty.

Reid, Walter, and Kenton Miller. **Keeping Options Alive: The Scientific Basis for Conserving Biodiversity.** Washington, DC: World Resources Institute, 1989. 130 pages. ISBN 0-915825-41-4

This report—atlaslike but concise—is a primer on biodiversity. It addresses concerns such as where the highest concentration of species live, how seriously biodiversity is threatened, the state of genetic diversity in agriculture, and what the best tools are for conservation of biodiversity.

The World Conservation Union, United Nations Environment Program, and World Wide Fund for Nature. **Caring for the Earth: A Strategy for Sustainable Living.** Gland, Switzerland: IUCN, UNEP, and WWF, 1991. 228 pages. ISBN 2-8317-0074-4

Nine principles, designed to help individuals, families, organizations, and nations establish societies that harmonize environmental conservation with economic development, form the base of this guide. Actions related to each principle are suggested, and further actions are recommended for different human activities (such as energy, farming, and business) or in the management of different biomes (such as forests, freshwater environments, and seas). The principles and the recommended actions are listed in the Statistics, Illustrations, and Documents chapter of this book.

Worldwatch Institute. **State of the World 1997.** New York: W. W. Norton, 1997. 229 pages. ISBN 0-393-04008-9

This annual publication tackles issues that the Worldwatch Institute finds the most pressing global challenges of the year, those that must be addressed along society's path toward sustainability. The *State of the World* series is consulted widely by corporate and government decision makers; the table of contents to each edition is a preview of issues that soon will become household concerns. The 1997 edition includes chapters on preserving global cropland, tracking climate change, cataloging nature's services, the hole in the ozone layer, and economic subsidies to environmentally damaging industries.

Anthologies and Conference Proceedings

Bailey, Ronald, editor. **The True State of the Planet.** New York: Free Press, 1995. 472 pages. ISBN 0-02-874010-6.

The contributors to this volume seek to change the focus of the conservation movement. They hold that overpopulation, food, global warming, and pesticide overuse are not the real problems threatening us; overfishing, lack of fresh water, and pollution in the developing world are much more important. The chapter on biodiversity introduces the topic and supports the position that people in developing countries know their own needs better than conservationists from wealthy nations. Sustainable development, says the chapter's author, is a better way to conserve biodiversity than highly technical preservationist techniques.

Potter, Christopher, Joel I. Cohen, and Dianne Janczewski, editors. **Perspectives on Biodiversity: Case Studies of Genetic Resource Conservation and Development.** Washington, DC: American Association for the Advancement of Science, 1993. 245 pages. ISBN 0-87168-512-4

This book focuses on the conservation of genes as the base of biodiversity. Case studies are described for genetic conservation in particular sites: agroecosystems, fisheries, wildlife reserves, and managed forests—all of these in the Americas, Africa, and Indonesia. The best techniques for conservation are culled from the studies and proposed for wider use.

Sandlund, O. T., K. Hindar, and A. H. D. Brown, editors. **Conservation of Biodiversity for Sustainable Development.** Oslo, Norway: Scandinavian University Press, 1992. 324 pages. ISBN 82-00-21508-3

The 1990 International Conference on the Conservation of Genetic Resources for Sustainable Development held in Røros, Norway, was the forum for the presentation of these 20 papers, most of them technical and very specific, on topics such as genetic resources and erosion, *ex-situ* conservation strategies, and biotechnology. Two of the more accessible chapters are written on a topic more often addressed by sociologists than biologists: inequities between the South, which has so much of the world's biodiversity, and the North, which has already destroyed its own and therefore is very eager to exploit what's left in the South. Tropical biologist Daniel Janzen describes the mission of Costa Rica's INBio (see Organizations), and Vandana Shiva of India (see the Biographical Sketches chapter) decries ignorance by scientists and development professionals of the native science and technology developed over millennia by indigenous peoples. The book concludes with an essay called "Sustainability! The Integral Approach" by noted philosopher Arne Naess.

Shiva, Vandana, Patrick Anderson, Heffa Schücking, Andrew Gray, Larry Lohmann, and David Cooper. **Biodiversity: Social and Ecological Perspectives.** London and Penang, Malaysia: Zed Books Ltd. and World Rainforest Movement, 1991. 123 pages. ISBN 1-85649-053-X

The contributors to this book contend that the prevailing approach to biodiversity conservation by scientific and conservation establishments is biased and flawed: "It is as if the mind is in the North, the matter is in the South; the solution is in the North, the problems in the South" (p. 7). The biodiversity crisis in poor, tropical countries can be traced back to the North, the authors charge, to unwise development projects and movements such as the Green Revolution, which robbed traditional farmers of their biodiverse agricultural methods. Topics covered include biotechnology, deleterious effects on indigenous peoples by conservation projects, Thai conservation strategies, and the need of poor countries to retain the rights to the genes found in their ecological communities.

Silver, Cheryl Simon, and Ruth S. DeFries. **One Earth, One Future: Our Changing Global Environment.** Washington, DC: National Academy of Sciences, 1990. 196 pages. ISBN 0-309-04141-4

Based on papers presented at the National Academy of Science's Forum on Global Change and Our Common Future, this book contains sections about the major ways that humankind is affecting the global environment. The changes we (and every other species on earth) will have to deal with include those from global warming, the hole in the ozone layer and the ultraviolet radiation it lets through, disappearing forests, and acid rain. The text includes enough background information to make it accessible and user-friendly. One of the appendices, a sort of executive summary of the book, was sent by the National Academy of Science as a "Letter on Global Environmental Change" to then-President-elect George Bush.

Solbrig, O. T., H. M. van Emden, and P. G. W. J. van Oordt, editors. **Biodiversity and Global Change.** Oxon, UK: CAB International, 1994. 227 pages. ISBN 0-85198-931-4

The International Union of Biological Sciences (IUBS) held a symposium in September 1991 on the effects of global change on biodiversity, and this volume contains the proceedings. The greenhouse effect (global warming due to increased amounts of carbon dioxide trapped in the atmosphere), habitat destruction, and releases of harmful chemicals such as DDT are all examples of global change. The chapters assess the effects of these phenomena on biodiversity overall, on microorganisms, and on marine and terrestrial systems. Sections on management and theory are included. This work is fairly technical in character.

Wilson, E. O., and Frances M. Peter, editors. **Biodiversity.** Washington, DC: National Academy Press, 1988. 521 pages. ISBN 0-309-03739-5

This volume is a collection of the papers presented at the 1986 National Forum on Biodiversity sponsored by the National Research Council of the National Academy of Sciences and the Smithsonian Institution. The 57 essays, all written by experts in their fields, are divided by theme (Challenges to the Preservation of Biodiversity, Human Dependence on Biological Diversity,

Restoration Ecology, etc.) and reflect scientific, conservationist, agricultural, philosophical, economic, and political perspectives on biodiversity.

Ethics and Biodiversity

Hamilton, Lawrence S., and Helen F. Takeuchi. **Ethics, Religion and Biodiversity: Relations between Conservation and Cultural Values.** Cambridge, UK: White Horse Press, 1993. 218 pages. ISBN 1-874267-09-X

As the title suggests, this collection of papers addresses ethical and religious considerations of biodiversity among cultures such as those of the Australian aborigines, the Xishuangbanna Dai community of southwest China, and the island people of Pohnpei, Micronesia, and native Hawaii. These peoples have revered their biodiversity and protected it by customs and traditional law. Other chapters reveal profoundly pro-biodiversity biases in the Bible and Buddhist dharma. Finally, a special overview provides the blueprint for conservation of biodiversity that can be constructed from the fields of ethics, culture, and religious studies.

Hannum, Hildegarde, editor. **People, Land and Community: Collected E. F. Schumacher Lectures.** New Haven, CT: Yale University Press, 1997. 328 pages. ISBN 0-300-07173-6

This collection of lectures by a diverse group of thinkers is dedicated to E. F. Schumacher's vision of human ecology (see E. F. Schumacher Society in the Organizations chapter). Together they propose a transformation of humanity's current anti-environmental tendencies into a new attitude and lifestyle that considers the health of all species on earth. The progression of sections in this book tells the story: (1) Beyond a Legacy of Domination, (2) Toward Decentralism and Community Revitalization, (3) Toward a New Era in Human-Earth Relations.

Kellert, Stephen R. **The Value of Life: Biological Diversity and Human Society.** Washington, DC: Island Press, 1996. 263 pages. ISBN 1-55963-318-2

For 20 years Stephen Kellert, a social ecologist, has been researching how humans value nature. This book collects his findings and thoughts. He classifies human attitudes toward nature into nine categories (utilitarian, naturalistic, ecologistic-scientific, aesthetic, symbolic, dominionistic, humanistic, moralistic, and negativistic) and shows how and why these attitudes manifest in various periods of history, among different ethnicities, and in the work of selected professions. Kellert concludes that humanity's psychological and socioeconomic well-being is contingent upon the preservation of biodiversity.

Kellert, Stephen R., and Edward O. Wilson, editors. **The Biophilia Hypothesis.** Washington, DC: Island Press, 1993. 484 pages. ISBN 1-55963-148-1

This volume responds to E. O. Wilson's *Biophilia* (1984; see below), which hypothesized that human beings have an innate fascination with other life-forms. Here scholars from a panorama of specialties write about their own related work. A psychiatrist and a researcher in behavior disorders describe an experiment at a residential school for children with learning and behavior problems. Some children were sent on an outdoor adventure and challenge-type recreational program, and others were allowed to adopt a pet on a school farm. The children with pets showed remarkable improvement in comparison to the others, evidence for biophilia-based therapy. Other chapters discuss native peoples' traditional knowledge of the species around them. New Guinean forest people distinguish all the species in their habitat recognized by science and actually differentiate between species that closely resemble each other and sometimes confuse taxonomists. And the species they differentiate are not only the ones they hunt or gather. This chapter, written by evolutionary biologist Jared Diamond, claims that their keen taxonomic skills have developed through close observation of ecological interactions between the species they cull for food and others that resemble them or that live in close contact with them. The 15 chapters in this collection prove that biophilia is a complex topic with many social ramifications.

Leopold, Aldo. **A Sand County Almanac and Sketches Here and There.** New York: Oxford University Press, 1949. 226 pages. ISBN 0-19-500777-8

Reading this book almost 50 years after it was published, one realizes how far ahead of his time Leopold was in preaching a heart-based conservation ethic. The first section, *A Sand County Almanac,* charts the passage of seasons at Leopold's Wisconsin sand farm. This is nature writing at its best, touching on tragic losses of rare prairie species like the Silphium, of which the last individual in Leopold's county, surviving for decades in an old pioneer cemetery, was finally conquered by highway mowers. In the last chapter, "The Land Ethic" traces the roots of ethics as a civilizing tool and proposes that we treat other species and the land we share with them with the same respect we purport to have for one another.

Wilson, E. O. **Biophilia.** Cambridge, MA: Harvard University Press, 1984. 157 pages. ISBN 0-674-07441-6

This was E. O. Wilson's first book of a personal nature; he followed it with the lyrical volumes *The Naturalist* and *In Search of Nature* described below. His leadership in the development of the field of ecology during the latter half of this century, coupled with his eloquence, make these books insightful, accessible introductions to the field. *Biophilia* ventures into a question most biologists would rather leave to anthropologists: how humans' feelings for nature have developed. In the book's prologue he writes: "We learn to distinguish life from the inanimate and move toward it like moths to a porchlight. Our existence depends on this propensity, our spirit is woven from it, hope rises on its currents." His identification of this tendency, and the urgency he assigned it, helped make the biophilia hypothesis into an issue later taken up by many professions (see *The Biophilia Hypothesis* above).

———. **The Naturalist.** Washington, DC: Island Press, 1994. 380 pages. ISBN 1-55963-288-7

These are the memoirs, both personal and professional, of the noted Harvard University biologist. The book traces Wilson's development from a shy Florida boy who spent his free time wandering through the natural areas near his home into a professional scientist. The half of the book devoted to his adult life is an intellectual history of the fields he has been interested in: population biology, biogeography, sociobiology, and biological diversity. He describes and praises the personalities of the scientists he has collaborated and competed with and explains the significance of their work.

———. **In Search of Nature.** Washington, DC: Island Press, 1996. 214 pages. ISBN 155-96-32151

E. O. Wilson once again articulates the value of biodiversity in a series of essays that reflect upon the questions that arise in his other books, among them: What is the genetic base of human behavior? If people can be said to partake in biophilia, then why are we so rapidly destroying our natural environment? What do insects do to keep the earth inhabitable for everyone else?

Biodiversity Detailed: Focus on Specific Ecological Communities, Taxonomic Groups, or Scientific Questions

Barrientos, Zaidett, and Julián Monge-Nájera. **The Biodiversity of Costa Rica.** Translations by Mary Jane Curry. Heredia, Costa Rica: INBio, 1995. 215 pages. ISBN 9968-702-02-1

A collection of color photographs and vignettes about the species-rich ecosystems of Costa Rica, this book was published by the National Biodiversity Institute (INBio). The 39 chapters are based on studies by Costa Rican and foreign biologists and are written in a readable style for ecotourists, the visitors who come not only to enjoy Costa Rica's natural resources but to learn something about them, too. Included in this entertaining, bilingual paperback are portraits of individual species and some of the fascinating interactions between them, as well as local efforts to save endangered species. This book is not widely available in the United States but can be ordered from INBio at Apartado 22-3100 Santo Domingo de Heredia, Costa Rica; fax: 506-244-2816.

Facklam, Howard, and Margery Facklam. **Plants: Extinction or Survival.** Hillside, NJ: Enslow Publishers, 1990. 96 pages. ISBN 0-89490-248-2

Written for secondary school researchers, this general work provides overviews of botany, agriculture, ethnobotany, biotechnology, extinction, seed banks, and medicinal plants. A good first source for a paper on a flora-related topic.

Forsyth, Adrian, and Ken Miyata. **Tropical Nature: Life and Death in the Rain Forests of Central and South America.** New York: Simon and Schuster, 1984. 249 pages. ISBN 0-684-18710-8

Clear in its preface that tropical nature includes much more than rain forests, this book nonetheless focuses on the biodiversity of the crème de la crème of biomes, the rain forest. The book's two authors, superb writers, cull their descriptions from their many combined years in the field, providing an insider's view to a place considered alien by most U.S. readers. Chapter-long essays cover topics such as ants, medicinal compounds, pollination, and mimicry. In the years since it was published, this book has become known as a valuable primer for anyone interested in tropical ecology.

Goodman, Susan E. **Bats, Bugs, and Biodiversity.** New York: Atheneum Books for Young Readers, 1995. 46 pages. ISBN 0-689-31943-6

A chronicle of the summertime voyage by a group of Michigan middle school students to the Amazon Basin. They spent the year prior to the trip studying the rain forest and raising money for their travels. While in the rain forest, they take science workshops about forest mechanisms, learn about jungle medicines, meet *ribereño* (riverside-dwelling) children to play, dance, and exchange gifts, learn about the kids' responsibilities in maintaining a jungle household, and barter with a local tribe. By departure time, the students have a new appreciation for rain forest conservation issues. They have learned that the baby sloth they bought in a market in order to free it in the forest probably would not survive without a mother as guide, and their buying it may encourage poachers to capture more. Any successful attempt at "saving the rain forest" must allow the rain forest's native inhabitants to live in a dignified, healthy way. Although poignant and informative for all readers, this book is written for young readers. It is illustrated with many beautiful photos.

Gould, Stephen Jay. **Eight Little Piggies: Reflections in Natural History.** New York: W. W. Norton, 1993. 479 pages. ISBN 0-393-03416-X

The latest in an ongoing series of Gould's collections of essays on a wide range of natural history topics, this volume begins with the sad story of the *Partula* land snails of Tahiti, 16 species of which had evolved in the steep volcanic valleys of the island of Moorea. They were a treasure to evolutionists, who used them to discuss whether the physical differences in the species were due to environmental factors or mutations that dominated due to isolation from other populations. Just as the debate was becoming most intense, connoisseurs of escargot (edible snails) introduced a fleshy African snail onto the island. Unfortunately, in addition to being juicy and delicious, this snail (of the *Achatina* genus) is cannibalistic, tracking and devouring other snails. In fewer than ten years after its introduction, it had eaten the very last *Partula*. Other essays mourn unnecessary extinctions caused by human intervention and describe cases in which humans probably had no hand at all. Evolution is given lots of coverage as well. Throughout this book and the others in this series, Gould's tales entertain, inform, and warn the general reader.

Hubbell, Sue. **Broadsides from the Other Orders: A Book of Bugs.** New York: Random House, 1993. 276 pages. ISBN 0-679-75300-1

This celebration of bugs is written in an informative and entertaining style. The author ventures off to assist different entomologists with their tasks, counting butterflies or discovering the virtues of gnats, for example. Each chapter is devoted to a different order: Lepidoptera (butterflies), Diptera (gnats and flies), Coleoptera (beetles), Opiliones (daddy longlegs), and others.

Liptak, Karen. **Saving Our Wetlands and Their Wildlife.** New York: Franklin Watts, 1991. 63 pages. ISBN 0-531-20092-2

Written for young readers, concise and interesting with plenty of color photos, this slim volume includes descriptions of the different types of salt- and freshwater wetlands: swamps, marshes, bogs, bottomlands, prairie potholes, and riparian wetlands. Also examined are their important roles as breeding grounds for fish and birds, stopover points for migrating flocks, sinks for water storage, sponges for flood control, and filters for contaminants. Unfortunately, they are disappearing from the U.S. landscape at a rate of a 1.5 million acres per year—a dramatic decrease given that only 95 million of the original 200 million acres of wetland remain. Almost half the book is devoted to comeback stories of communities learning to love their wetlands and the animals associated with them. It concludes with a list of public wetlands one can explore and what to wear and bring on the trip.

Pimm, Stuart L. **The Balance of Nature? Ecological Issues in the Conservation of Species and Communities.** Chicago: University of Chicago Press, 1992. 434 pages. ISBN 0-226-66829-0

Are simple ecological communities (such as monocropped agricultural tracts) less stable than more diverse or complex ones (such as an intact grassland)? Part of the difficulty scientists have had answering this question comes with the word *stable*. Stability can be variously interpreted by biologists as *mathematical stability*, where communities have a natural equilibrium and will return to it after a disturbance; *resilience*, a measure of how quickly the community in question returns to its equilibrium after a disturbance; *persistence*, or how long a community can hold out before it surrenders to an invasion of some sort; *resistance*, the consequences for the overall community of a shift in its species composition; and *variability*, or how a community varies. Another common point of confusion is the definition of complexity. It can mean the number of species in a community, the abundance of a species within a community, or the degree of connectedness the species have to each other through their food web. After elucidating these points of confusion, the author examines each subdefinition and how it contributes to the whole picture. This book is one of the more technical works listed in this chapter.

Smith, Nigel J. H., J. T. Williams, Donald L. Plucknett, and Jennifer P. Talbot. **Tropical Forests and Their Crops.** Ithaca, NY: Cornell University Press, 1992. 568 pages. ISBN 0-8014-2771-1

This reference book surveys tropical forest crops that provide people with beverages (coffee, cacao, cupuaçu), fruit (mango, citrus, pineapple, avocado, guava, papaya, sapodilla, passion fruit, durian, rambutan, and more), starch (breadfruit, plantains), oil (palms), resins (pines, balsams), fuelwood and fodder (bamboo), spices (clove, cinnamon, vanilla, annato), and nuts (cashew, Brazil nuts, macadamia). Each crop's entry contains a photo of the product and considers its socioeconomic importance, including how it is used locally and internationally, how it is cultivated, and the lore that has developed around it locally. Also mentioned is how well protected the crops' genetic resources are: Is the forest where wild cousins grow in danger of disappearing? Has germ plasm been preserved in gene banks and is it used by breeders? The resources described in this book are convincing evidence of the importance of tropical crops.

Terborgh, John. **Diversity and the Tropical Rain Forest.** New York: Scientific American Library, 1992. 242 pages. ISBN 0-7167-5030-9

This book addresses a very specific topic, which sets it apart from most books on biodiversity or the rain forest. Since the author limits his focus to why tropical rain forests are as diverse as they are, he has time to address some very specialized issues. One chapter focuses on sunlight and stratification—how species benefit from the differing amounts of sunlight that reach them as it filters down through the canopy, and how tropical tree crowns have evolved to be flat in order to take advantage of the 90-degree angle at which the sun hits the earth in the tropics. (In the higher latitudes, where the sun is lower in the sky during half the year, tree crowns are more conical in shape.) Another chapter is devoted to convergence, "the theory that life forms in similar environments independently evolve similar adaptations"—either physical traits, like the red monkeys of Asia, Africa, and the New World, or behaviors, like birds picking insects out of crannies in bark or snapping them up in flight. The author, a U.S. tropical biologist who has spent many years managing a research station in Peru's Amazonian Basin, grapples with erudite topics in a way that makes them accessible to the lay reader.

Biodiversity in Crisis

Ackerman, Diane. **The Rarest of the Rare: Vanishing Animals, Timeless Worlds.** New York: Random House, 1995. 186 pages. ISBN 0-679-40346-9

Ackerman writes in her introduction: "For the past years I've been traveling to see some of the rarest animals and ecosystems. As a member of the species responsible for their downfall, I feel an urgent need to witness and celebrate them before they vanish." This book, resulting from her series of pilgrimages, contains chapters on the monk seal, the short-tailed albatross, the golden lion tamarin, and other endangered species. Each chapter takes her to an exotic habitat, accompanied by biologists who study and strive to protect their chosen species. Ackerman mixes biographies of her human companions with the biology of her nonhuman companions, and her compelling, poetic writing style makes this book hard to put down. It's a good nontraditional supplement to any research project on endangered species.

Collar, N. J. **Birds to Watch 2: The World List of Threatened Birds.** Birdlife Conservation Series, no. 4. Norwich, UK: Birdlife International, 1994. 407 pages. ISBN 1-56098-528-3

Dictionary-like, this book is a listing of extinct, endangered, vulnerable, and rare birds worldwide. Compiled by the conservation organization Birdlife International, it serves as a base for IUCN's *Red Data Book* on birds (see introduction to this chapter). Information about each species, including range, sightings, and type of threat to its survival, are provided in the main body of the book. Appendices list threatened birds by country or territory, countries or territories in order of number of threatened species, and species by threat categories.

Ehrlich, Paul, and Anne Ehrlich. **Extinction: The Causes and Consequences of the Disappearance of Species.** New York: Random House, 1981. 306 pages. ISBN 0-394-51312-6

The Ehrlichs are pessimistic about the outcome of the human onslaught on the rest of the world. But this book is neither glum nor dreary; it is a very readable examination of extinction in the past, present, and future, peppered with anecdotes from the writers' decades of research around the world. Included, for example, are theories about why water snakes of the same species have dark bands if they live on the shores of Lake Erie and no bands if they live on the lake's limestone islands, and the story of how an entire population of a small blue butterfly that feeds on a single species of flower in the Colorado Rockies was wiped out by a late spring freeze. The book also provides useful background information on the evolutionary processes that brought about the biodiversity now so quickly disappearing and ends with some hopeful scenarios of a world in much less danger.

Eldridge, Niles. **Dominion.** Berkeley and Los Angeles: University of California Press, 1995. 190 pages. ISBN 0-520-20845-5

Author Niles Eldridge, a well-known evolutionary biologist, uses an essay format to trace human interactions with the environment from prehistory to the present. His examination of archaeological sites has convinced him that humanity's exploitation of and disregard for the health of the natural environment started with the development of agriculture 10,000 years ago. Exploitation will not end, he predicts, until the human population declines and wealthy nations reduce their overconsumption of resources.

Erickson, John. **Dying Planet: The Extinction of Species.** Blue Ridge Summit, PA: TAB Books, 1991. 188 pages. ISBN 0-8306-6726-1

This is an introductory primer to the beginnings of life-forms and how extinctions have cut them off mid-stride. The information is condensed into a concise, well-organized series of illustrated chapters that make it a useful crash course: The Origin of Life, The History of Life, The Major Extinctions, Evolution of Species, etc.

Kaufman, Les, and Kennoth Mallory, editors. **The Last Extinction.** 2d. ed. Cambridge, MA: MIT Press, 1993. 250 pages. ISBN 0-262-61089-2

This book is an updated second edition of one of the first books addressing species and habitat loss. It was designed to set the record straight on some misconceptions about the mass extinction wave currently occurring and inspire public response to it. The chapters are brief and lucid and provide good overviews of their topics: mass extinctions of the past, current extinctions in North America and the Amazon, dangers faced by whales, and captive breeding efforts.

Raup, David M. **Extinction, Bad Genes or Bad Luck?** New York: W. W. Norton, 1991. 210 pages. ISBN 0-393-03008-3

Paleontologist David Raup examines the possible causes of the "Big Five"—the five mass extinctions of the Paleo-zoic and Mesozoic eras. Chapters are dedicated to biological causes of extinction (small populations unable to rebound from catastrophe, too many species inhabiting an area too small to support them, and competition between species) and physical causes (sea level and climate change, volcanism, and meteorites).

Shiva, Vandana. **Monocultures of the Mind: Perspectives on Biodiversity and Biotechnology.** London and Penang, Malaysia: Zed Books Ltd. and Third World Network, 1993. 184 pages. ISBN 1-85649-217-6

The five essays in this volume, culled from Shiva's writings of the previous decade, focus on how the disappearance of biodiversity is rooted in the loss of traditional knowledge and practices. The essays touch on Shiva's areas of expertise: the Green Revolution, ongoing North-South conflicts in biodiversity management, the 1992 Biodiversity Convention, and the need to conserve what she terms "the seed and the spinning wheel"—indigenous science and technology.

Ward, Peter. **The End of Evolution: On Mass Extinctions and the Preservation of Biodiversity.** New York: Bantam Books, 1994. 302 pages. ISBN 0-553-08812-2

Two major extinctions of ancient history are examined in this book. The first occurred at the end of the Permian period 250 million years ago, the second at the boundary of the Cretaceous and Tertiary periods 65 million years ago. Both of these crises were marked by the extinction of more than 90 percent of the species then existing. The author equates these periods to the extinction spasm we are currently experiencing and warns that if this crisis is like those of the past, biodiversity will not rise to the level we currently enjoy for 10 million years.

Weiner, Jonathon. **The Next One Hundred Years.** New York: Bantam Books, 1990. 312 pages. ISBN 0-553-05744-8

This book muses on the threats to life on earth that have arisen due to humankind's interference with nature and describes theories and experiments that attempt to preserve biodiversity. Special attention is given to the greenhouse effect, the hole in the ozone layer, the Gaia Hypothesis, and the human population explosion. This well-written book is most useful when read in its entirety. It reads like a very long essay—discussions commence, are put aside for a chapter or two, then later resume at another point.

Indigenous Peoples and Biodiversity

Foster, Nelson, and Linda S. Cordell. **Chilies to Chocolate: Food the Americas Gave the World.** Tuscon: University of Arizona Press, 1992. 191 pages. ISBN 0-8165-1324-4

This multifaceted book covers the history, economics, botany, and nutritional value of the cornucopia of foods native to the Americas. Fruits, vegetables, nuts, and grains such as blueberries, cranberries, black walnuts, wild rice, vanilla, corn, beans, amaranth pumpkins, squash, potatoes, and tomatoes were "discovered" by European explorers and taken back to their homeland. Initially they were received with confusion and disgust, but little by little some of them claimed positions of importance within European cuisine. A parallel focus of the book is the tragic shift in American agriculture away from these nutritional native crops to the foods Europeans were more familiar with. Currently, Native American activists are attempting to rekindle interest in traditional cultivation methods and crops, and organic farmers are reviving old grains like amaranth and quinoa.

Inter Press Service, compiler. **Story Earth: Native Voices on the Environment.** San Francisco, CA: Mercury House, 1993. 200 pages. ISBN 1-56279-035-8

Indigenous views, stories, and visions of nature from North, Central, and South America, from Sri Lanka, New Guinea, India,

Kenya, Lesotho, Finland, and many other areas of the world are presented in this diverse anthology.

Nabhan, Gary. **Enduring Seeds: Native American Agriculture and Wild Plant Conservation.** See page 191.

Palmer, Paula, Juanita Sánchez, and Gloria Mayorga. **Taking Care of Sibö's Gifts.** San José, Costa Rica: Editorama, 1993. 98 pages. ISBN 9977-88-019-0

Subtitled *An Environmental Treatise from Costa Rica's KéköLdi Indigenous Reserve,* this book was written by two KéköLdi women and a North American sociologist collaborator. The KéköLdi reserve is situated in the southern Caribbean coastal and mountainous region of Costa Rica. With fewer than 200 residents, it is one of the tiniest of the country's 21 small indigenous reserves. Yet the KéköLdi are an inspiring model for how a tropical people can sustainably use their rain forest resources. Everything about their way of life reflects their respect for and dependence on the gifts of Sibö (the Creator), from their cosmology to their domestic lifestyle, to how they have dealt with the influx of the modern world. The book describes their social organization, language, creation stories, Sibö's laws about use of natural resources, customs, and the role of healers. Tribe members' stories are compiled oral-history style, and there is lots of nuts-and-bolts information about reserve boundaries, how they build their homes, and how their iguana-raising project got started and is proceeding. An appendix offered by a KéköLdi shaman shares medicinal uses of plants. This is a very useful primary source for anyone interested in true sustainability. Not readily available in bookstores, this book can be ordered in English or Spanish for $12 from Paula Palmer, P.O. Box 7490, Boulder, CO 80306; (303) 444-0306, fax: (303) 449-9794; globresponse@igc.apc.org.

Plotkin, Mark J. **Tales of a Shaman's Apprentice.** New York: Viking, 1993. 319 pages. ISBN 0-670-83137-9

Nine separate quests into Amazonia are documented here by Plotkin, a well-known ethnobotanist who studies how tribal shamans, or medicine men, use rain forest plants and animals as medicine, poison, and hallucinogens. Plotkin's openness allows him to live like his hosts, and his adventurous character leads him to experiment with potent compounds. He is not the type of researcher who uses native peoples just as sources of information. His sensitivity and respect toward his tribal hosts are apparent throughout the book, and he has made great contributions to helping conserve both rain forest resources and indigenous knowledge of them. He discusses issues of intellectual property and sustainable harvesting of rainforest resources. Ethnographic anecdotes are so artfully combined with descriptions of lush tropical nature that this book is memorable not only for the important information it contains but also as a spellbinding travelogue.

Wenz, Peter S. **Nature's Keeper.** Philadelphia, PA: Temple University Press, 1996. 207 pages. ISBN 1-56639-427-9

Wenz's work identifies a central difference between the dominant European-based culture and that of indigenous peoples. For Europeans, natural resources exist for the benefit of humans; for indigenous peoples, nature is respected in and of itself. Examples of the indigenous ethic are taken from speeches by indigenous participants in the World Uranium Hearing held in 1992 in Salzburg, Austria. They reveal cultures that revere all species and place human beings in the same family as birds, mammals, and plants. To support his thesis about European culture, the author deconstructs the four most important ideas, movements, and practices that form its historical cornerstones: Christianity, commercialism, industrialism, and modern bureaucracy. For each of these he discloses a five-step pattern: (1) people see themselves as separate from nature; (2) people perceive themselves as threatened by nature; (3) people subdue nature, but this actually increases the danger they are in; (4) since people are in danger, they concentrate force; and (5) they use this force against other people. The conclusion of the book is that people of European-based cultures must adopt some of the perspectives of indigenous people in order to develop a more equitable relationship with other people and other species on earth.

Agriculture and Biodiversity

Ausubel, Kenny. **Seeds of Change: The Living Treasure.** San Francisco, CA: HarperCollins, 1994. 232 pages. ISBN 0-06-250008-2

This is the history of Seeds of Change, a supplier of organic seeds for heirloom crops and rare varieties of fruits, vegetables, grains, and legumes. Vibrant personalities joined to form this company, and their passion for seeds and gardening, vividly described in the book, are infectious—if you are not a gardener, you will want to become one. The history of the mainstream seed business is also discussed, showing how companies moved increasingly in the second half of this century toward pesticide-dependent hybrids selected more for productivity and attractiveness than for environmental or taste advantages. Because their plants were hybrids and didn't produce viable seeds, gardeners using them had to abandon their traditional practice of saving seeds for the next year. They, and more importantly, large-scale farmers become hooked into dependency on seed companies and the chemical inputs required to grow their crops. This book is lucid, easy to read, and hard to put down.

Buchmann, Stephen L., and Gary Nabhan. **The Forgotten Pollinators.** Washington, DC: Island Press, 1996. 292 pages. ISBN 1-55963-352-2

Flowering plants and the animals that pollinate them together make up two-thirds of the species so far identified on earth. They have coevolved to form intimate, productive relationships. One in every three bites we take is made possible by the complex, mutually beneficial interactions between plant and pollinator. Yet due to shrinking natural habitat, flowering plants and their pollinators are in decline. This book is one of the products of the Forgotten Pollinators Campaign, sponsored by the Arizona-Sonora Desert Museum with support from several foundations. The book celebrates the diversity of pollinators, describes the extent of the crisis, and provides suggestions for their restoration. (See also the Forgotten Pollinators Campaign in the Statistics, Illustrations, and Documents chapter.)

Carson, Rachel. **Silent Spring.** Boston: Houghton Mifflin, 1962. 368 pages. ISBN 0-395-45389-5

When this book was published, it sounded a warning: overuse of chemical pesticides will destroy ecosystems and trigger major cancer outbreaks. Carson's poetic style coupled with her scientific knowledge alerted Americans to contamination of their food supply (even baby food and mother's milk contained intolerable quantities of DDT residues) and to the vicious cycle whereby pests evolved resistance to pesticides. The title refers to the worst-case scenario: songbirds, poisoned by pesticide-sprayed insects, would never again enliven spring with their music. *Silent Spring* incited the people of the United States to demand important policy changes toward dangerous and unnecessary chemical pesticides. Unfortunately, however, those laws have not curbed the use of poisons, and much of the horror depicted by the book still occurs.

Nabhan, Gary. **Enduring Seeds: Native American Agriculture and Wild Plant Conservation.** San Francisco, CA: North Point Press, 1989. 225 pages. ISBN 0-86547-343-9

Ethnobiologist and storyteller Gary Nabhan explores the riches of preconquest indigenous agriculture, mourns its demise and the ensuing erosion of genetic diversity, and describes the small-scale but enthusiastic revival of native agriculture. Although Nabhan focuses on the practices and crops of the Sonoran peoples—he is based in Tuscon, Arizona—he also takes the reader to northern Minnesota and Wisconsin to observe a harvest of wild rice, to the Everglades, to the central Mexico site of a sumptuous garden planted in the 1400s for Aztec aristocrats to bask in the aromas of exotic flowering plants, and to many other locales.

Legal and Legislative Responses to the Biodiversity Crisis

Bean, Michael J., Sarah G. Fitzgerald, and Michael A. O'Connell. **Reconciling Conflicts under the Endangered Species Act.** Washington, DC: World Wildlife Fund, 1991. 109 pages. ISBN 0-89164-130-0

Listing a species as "endangered" under the U.S. Endangered Species Act (ESA) is, unfortunately, not enough to protect it from threats to its survival. Many endangered species inhabit only private land—land whose owners are not necessarily conservationists. Conservationists and local governments may use the ESA to force landowners to protect engangered species when the owners seek permission to develop their land—agreeing to help the species in question may be the condition for granting permission to build. This handbook describes the complex research and negotiation processes that are involved in hammering out an acceptable Habitat Conservation Plan that assures the health of the species in question. It includes four case studies where compromises have been worked out.

Fitzgerald, Sarah. **International Wildlife Trade: Whose Business Is It?** Washington, DC: World Wildlife Fund, 1989. 459 pages. ISBN 0-942635-10-8

International trafficking in endangered wildlife is big business, but because of it takes place "underground," it is very hard to control.

The author of this book estimates that international trade in wildlife and wildlife products, though not a principal factor in the current mass extinctions, affects 40 percent of all vertebrate species. This book includes a short introduction to the Convention on International Trade in Endangered Species (CITES) and its predecessor treaties, followed by chapters on species particularly affected by international trade, including mammals (bears, cats, elephants, kangaroos, musk deer, primates, rhinos, vicuñas, otters, seals, walruses, and whales), birds (parrots, raptors, songbirds, hummingbirds, and those prized for their feathers), reptiles (crocodiles, lizards, snakes, and turtles), butterflies, corals, ornamental fish, amphibians, spiders, and many types of live plants. Grisly photos are reminders of the nature of this business.

Grubb, Michael, Matthias Koch, Koy Thomson, Abby Munson, and Francis Sullivan. **The "Earth Summit" Agreements: A Guide and Assessment.** London: Earthscan Publications Ltd, 1993. 180 pages. ISBN 1-85383-176-X

This handy guide provides a background to the 1992 Rio Summit negotiations, analyzes the summit itself, and predicts how each of the treaties that came out of the summit will be implemented. Chapters elucidate each of the agreements resulting from the summit: the Convention on Climate Change, the Convention on Biodiversity, the Convention on Environment and Development, Agenda 21, and the Forest Principles.

Stone, Christopher D. **The Gnat Is Older than Man: Global Environment and Human Agenda.** Princeton, NJ: Princeton University Press, 1993. 341 pages. ISBN 0-691-03250-5

This thought-provoking work approaches conservation from legal and economic perspectives. It brings up issues like cross-border contamination, in which pollution originates in one country but affects either another country or "the global commons" (oceans, the atmosphere). Assigning responsibility is important, but the greater task is designing mechanisms for compensating victims, cleaning up the mess, and preventing repeat occurrences. The author devotes a chapter to international treaties and assuring compliance. In his discussion of economics, Stone proposes a Global Commons Trust Fund whose coffers would be filled with taxes charged for all greenhouse gas emissions. It could fund atmospheric and ocean cleanups. Regarding conservation of biodiversity, the book describes mechanisms for conservation already in place: debt-for-nature swaps, transfer of appropriate technology, and the practice of making conservation a condition for economic assistance.

In-situ Conservation Approaches

Bonner, Raymond. **At the Hand of Man: Peril and Hope for Africa's Wildlife.** New York: Alfred A. Knopf, 1993. 324 pages. ISBN 0-679-40008-7

New York Times reporter Raymond Bonner explores in depth the reasons for the utter failure of many conservation policies in Africa. He focuses on the ivory ban that international conservation organizations forced through the 1989 Convention on International Trade in Endangered Species (CITES) conference, despite evidence from all over the African continent that such a ban was not necessary and in fact might threaten populations of the African elephant. The book presents a convincing argument that native peoples must be involved in conservation work in their countries for it to succeed, with two case studies of successful projects. In both examples, villagers who live next to wildlife reserves come to view wildlife not as a liability to their crops, livestock, and families but as a benefit from which they will reap a profit now and for years to come.

Bryant, Dirk, Daniel Nielson, Laura Tangley, and Nigel Sizer. **The Last Frontier Forests: Ecosystems and Economies on the Edge.** New York: World Resources Institute, 1997. 80 pages. ISBN 1-56973-198-5

Since there are no traditional frontiers left on earth, no limitless, unexplored stores of treasures like precious metals or timber, it is time to take stock of the riches that are left. This report redefines "frontier" as a place with a rich natural and cultural heritage and advocates a new stewardship of such valuable areas. The book includes many large maps of the world's existing forests, and the format allows for easy, systematic comparisons. Analyses of threats to remaining forests, and options for reversing them, are also provided.

Budiansky, Stephen. **Nature's Keepers: The New Science of Nature Management.** New York: Free Press, 1995. 310 pages. ISBN 0-02-904915-6

Budiansky offers a critique of traditional twentieth-century wildlands management. He bites into Smokey the Bear, charging that half a century of preventing fires has turned our forests into tinderboxes. In history, fires occurred frequently, set by lightning or by humans. Fires were thought to keep a forest young and healthy. But with decades of Forest Service antifire policies, flammable fallen trees and old growth have accumulated. One result of this management practice was the 1988 inferno in Yellowstone, which was much larger and hotter than any previously documented fire there. Budiansky also criticizes campaigns to save animals like deer, elk, feral horses, and burros, because without any means of population control (no hunting and few natural predators) they rapidly expand and eat through their own (and other animals') habitats. The real culprit, in Budiansky's view, is Western civilization's conception that nature will flourish if we only leave it alone. Budiansky argues that merely turning away is irresponsible, because the current problems were created by people in the first place and cannot be solved without human involvement. He pins his hopes on restoration ecology, or the recreation of lost habitats through lots of hard labor.

DiSilvestro, Roger L. **Reclaiming the Last Wild Places.** New York: John Wiley & Sons, 1993. 266 pages. ISBN 0-471-57244-6

Wildlands managers, biologists, and conservationists are realizing that the traditional approach of setting certain lands aside because they are of great natural value is not enough to assure their preservation. A national park or national forest cannot be cordoned off from its neighbors. These places are vulnerable to the effects of all the destructive activities that happen outside of their boundaries, such as the spraying of toxic chemicals on crops, golf courses, or suburban backyards; the dumping of industrial waste into rivers; and the killing of wild predators of livestock that stray from the protected area into rangeland. Wetland preserves are in the most danger because they are fed by rivers, often very contaminated ones, that originate in unprotected areas. The first chapter in this book is devoted to the tragic decline of the first national park set up to protect wilderness and wildlife rather than stunning geological features. In the 50 years since its establishment, Everglades National Park has become a virtual sewer, the most threatened national park in the United States. This sobering initiation precedes chapters entitled "The Invention and Overthrow of Wilderness," "Lands without Meaning," and "The Shattered Cradle." They provide historical information about the dawn of the North American movement to preserve wildlands and give status reports on national parks, national forests, and national wildlife refuges. Although it is full of bad news, this book is important for anyone interested in the realities of wildlife and wildlands conservation.

Grumbine, R. Edward. **Ghost Bears: Exploring the Biodiversity Crisis.** Washington, DC: Island Press, 1992. 294 pages. ISBN 1-55963-152-X

This book describes and critiques traditional ecosystem management, with a focus on the Greater North Cascades in the Pacific Northwest. Fragmented wilderness crisscrossed by roads and bald demarcation strips, with patches owned by various entities and exploited in different ways, cannot sustain wide-ranging animals like wolves, bears, and felines. The author, a conservation biologist, decries wildlands management based on "resourcism"—the ideology that wildlands should be protected only because they protect resources that humans might need to exploit later. Although he cannot predict when it will happen, Grumbine hopes that North America will lead the world in restoring its original biodiversity. For this to happen, three conditions must be met: individuals and institutions must internalize a biocentric ethic, there must be widespread acceptance of conservation biology's justification of interconnected wilderness areas, and decisions on land-use matters must be made with democratic participation at community and regional levels.

Guidelines on the Conservation of Medicinal Plants. Gland, Switzerland: IUCN, WHO, and WWF, 1993. 52 pages. ISBN 2-8317-0136-8

This short book outlines a comprehensive global protection plan for medicinal plants developed during worldwide conservation meetings in the late 1980s and early 1990s. The focus is on ensuring the conservation of medicinal plants, and the book provides guidelines for their sustainable use. It assigns responsibility to those with crucial roles in the task: agronomists, conservationists, ecologists, ethnobotanists, horticulturists, health policymakers, legal experts, park planners and managers, plant breeders, religious leaders, taxonomists, and traditional health practitioners.

Kramer, Randall, Carel van Schaik, and Julie Johnson. **Last Stand: Protected Areas and the Defense of Tropical Biodiversity.** New York: Oxford University Press, 1997. 242 pages. ISBN 0-19-509554-5

The editors of this volume decry the economic overtones given to recent conservation management strategies, claiming that biodiversity should be preserved for its intrinsic worth, not for its potential economic benefit to humans. They criticize the conservationist premise that indigenous peoples are the best caretakers of their wild homes, claiming that most indigenous groups can no longer live "sustainably" on their land since they have adopted the ways of the Western world. Contributors to this volume address topics such as the varying paradigms for conservation, land-use strategies, the politics of biodiversity, user rights, who should pay for conservation, and how to compensate people who preserve their land.

Krasemann, Stephen J., and Noel Grove. **The Nature Conservancy: Preserving Eden.** New York: Harry N. Abrams, 1992. 176 pages. ISBN 0-8109-3663-1

Dramatic, full-page photographs illustrate this book celebrating the contributions by the Nature Conservancy to the protection of biodiversity. The book contains descriptions of some of the Conservancy's many reserves, as well as anecdotal tidbits about their purchase and rehabilitation. Santa Cruz Island in California's Santa Barbara Channel, for example, was infested with feral sheep that had destroyed much of the island's flora. Environmentalists were forced to become uneasy but resolute hunters, eliminating the population sheep by sheep, and after the last was killed in 1987, the island's foliage began to recover its original diversity.

Lewis, Dale, and Nick Carter, editors. **Voices from Africa: Local Perspectives on Conservation.** Washington, DC: World Wildlife Fund, 1993. 216 pages. ISBN 0-89164-124-6

This book gives voice to African people involved in wildlife management and conservation on their own continent. Theirs are important voices to hear and contemplate, because so much of Africa's recent history is characterized by failed social development and conservation policies designed by outsiders who ignored the perspective of African people. Included are papers describing the most successful African wildlife management projects as well as the flawed ones. Even more interesting than these are the interviews with traditional chiefs, which are interspersed throughout the more technical, academic chapters. The chiefs describe traditional social structure in chiefdoms, how their beliefs lead to actions toward nature, flight from countryside to cities, and family planning.

Miller, Kenton. **Balancing the Scales: Guidelines for Increasing Biodiversity's Chances through Bioregional Management.** Washington, DC: World Resources Institute, 1996. 73 pages. ISBN 0-915825-85-6

The bioregional approach to conservation involves designing projects that are as large as an entire ecosystem or several neighboring ones, and engaging local people in the task of managing the area. This book studies nine cases where a bioregional management approach has been undertaken (La Amistad Biosphere in Costa Rica; Yellowstone; the Wadden Sea between the Netherlands, Germany, and Denmark; the Serengeti; the Great Barrier Reef; the Mediterranean Sea; the CAMPFIRE project in Zimbabwe; the United Kingdom's North York Moors National Park; and India's Hill Resource Management Program). Lessons are drawn from each example and synthesized. Although the intended audience is policymakers and managers, the introductory information on the bioregional approach and the case studies are of general interest.

Noss, Reed F., and Allen Y. Cooperrider. **Serving Nature's Legacy: Protecting and Restoring Biodiversity.** Washington, DC: Island Press, 1994. 416 pages. ISBN 1-55963-247-X

This comprehensive description of biodiversity and how best to manage it was written by two scientists from the forefront of the conservation biology field with support from Defenders of Wildlife (see Organizations). The first two chapters describe how biodiversity developed over the eons and how it is destroyed. The rest discuss its conservation: management of forests, rangelands, and aquatic systems; monitoring conservation projects; and barriers to conservation (philosophical, institutional, political, educational, and technical). The midsection of the book is aimed at management students or professionals, but the rest is quite accessible to a general reader.

Reid, Walter V., Sarah A. Laird, Carrie A. Meyer, Rodrigo Gámez, Ana Sittenfeld, Daniel H. Janzen, Michael A. Gollin, and Calestous Juma. **Biodiversity Prospecting: Using Genetic Resources for Sustainable Development.** New York: World Resources Institute, 1993. 342 pages. ISBN 0-9155825-89-9

Bioprospecting, a booming industry, entails a company (usually a pharmaceutical company) searching biodiverse ecosystems for useful compounds. It has great potential to fund *in-situ* conservation. This collection of articles focuses mainly on the experience of

Costa Rica's INBio (see Organizations), because it is the institution with the most experience in regulating bioprospecting. The authors caution about the harm that can be caused by unregulated bioprospecting, and they give suggestions for drawing up fair contract agreements between industry and host countries, establishing INBio-type institutions in other countries, granting permission for research and collection of specimens, intellectual property rights, and national policies.

Wallace, David Rains. **The Quetzal and the Macaw: The Story of Costa Rica's National Parks.** San Francisco, CA: Sierra Club Books, 1992. 222 pages. ISBN 0-87156-585-4

The government's conservation of wildlands in national parks and reserves is not an achievement to take for granted. It is a complicated and controversial task, especially in a country like Costa Rica, where there is little wildland left and what is left is threatened with deforestation by poor peasants, wealthy ranchers, and multinational agribusiness. This book tells an exciting story, recounting the adventures of Mario Boza and Alvaro Ugalde, the two Costa Ricans who envisioned and set into place that country's great national park system. Boza was the retiring intellectual author of the plan, and Ugalde was the gutsy front man who repeatedly risked his career and life to defend his beloved parks. Many other captivating personalities helped and hindered Boza and Ugalde, and they, too, are vividly depicted in the book. The author also traces the development of the ideology that determined how the parks were run—from fortresses jealously guarded from the parks' neighbors, all of whom were suspected of being poachers, to the current attempt to convert the parks' neighbors into allies in their protection. In the 25 years since the first two parks were demarcated, the country has set about 25 percent of its territory under some sort of protection. That feat is a challenge to the rest of the world, and reading about how Costa Ricans did it is worthwhile.

Ex-situ Conservation Approaches

Cohen, Daniel. **The Modern Ark: Saving Endangered Species.** New York: G. P. Putnam's Sons, 1995. 120 pages. ISBN 0-399-22442-4

Although their costliness and labor intensity make captive breeding projects controversial, they have brought some endangered species back from the brink of extinction. The Bronx Zoo saved the American bison around the turn of the century and repopulated the West with so many herds that the animal is no longer endangered. Written for a young adult readership, this book is a thorough and intelligent review of captive breeding efforts past and present. Included are case studies of attempts to replenish dwindling populations of the red wolf, California condor, black-footed ferret, peregrine falcon, golden tamarin, Arabian oryx, giant panda, and cheetah. The hills and valleys in each path are traced: health problems with the original breeders, public relations challenges with reintroduction of young animals into the wild, and the tragic impossibility of reintroducing a species into a wild that no longer exists.

Norton, Brian, Michael Hutchins, Elizabeth Stevens, and Terry Maples, editors. **Ethics on the Ark.** Washington, DC: Smithsonian Institution Press, 1995. 330 pages. ISBN 1-56098-515-1

The many contributors to this volume include zookeepers, philosophers, and genetics experts, all involved in the highly technical and conservation-oriented tasks of managing today's zoos. The book is divided into four sections. The first addresses the future of zoos. The second discusses what the target of protection should be: genes, individual animals, populations, entire species, or ecosystems. The third section, "Captive Breeding and Wild Populations," examines how animals are procured for zoos, what is done with "surplus" animals, and when reintroduction can be contemplated. The final chapter attacks the gnarly questions of how to care for captive animals and what types of research are ethical.

Restoration of Ecosystems

Falk, Donald A., Constance I. Millar, and Margaret Olwell, editors. **Restoring Diversity: Strategy for Reintroduction of Endangered Plants.** Washington, DC: Island Press, 1996. 505 pages. ISBN 1-55693-297-6

Since ecosystems everywhere have been damaged by human impact, restoration opportunities abound. Traditional ecology studies natural ecosystems in isolation, but increasingly, biologists find the need to intervene to save the subjects of their studies. This book is divided into five sections. The first defines when restoration is appropriate, the second lists some biological issues that determine the success or failure of restoration attempts, the third discusses the use of restoration in Endangered Species Act–mandated mitigation, the fourth provides several case studies, and the fifth gives guidelines for reintroducing endangered plants.

Mills, Stephanie. **In Service of the Wild: Restoring and Rehabilitating Damaged Land.** Boston: Beacon Press, 1995. 237 pages. ISBN 0-8070-8534-0

Rather than focusing on all the ways we are damaging our planet, this book visits places where ecological restoration efforts are healing it. The author begins at her personal 35-acre private restoration project and revisits it throughout the book. This is her own labor of love on the Leelanau Peninsula of Lower Michigan. Other restoration projects she describes include the North Branch Prairie Restoration Project (also discussed in William K. Stevens's book below); the Mattole Watershed Salmon Support Group in Northern California; and Auroville, an intentional community in the Tamil Nado state of southern India. All of these groups have made significant progress in helping their chosen ecosystems regain their health, and interestingly, the spiritual health of the people working on these projects has improved as well.

Stevens, William K. **Miracle under the Oaks.** New York: Pocket Books, 1995. 332 pages. ISBN 0-671-78042-5

Re-creating North America's long-lost ecosystems has become a mission for growing numbers of restoration pioneers. As organized groups of volunteer restorationists work, they discover, through much trial and error, what their target ecosystem needs to recover its original splendor: keystone plants, pollinators and seed dispersers, shade versus light, and species that have disappeared from the area but cling to abandoned lots and other ignored "wastelands." Once a few of these elements are in place, what seems to be the original ecosystem magically reappears. The case study receiving most coverage is the inspiring work of Steve Packard (see Biographical Sketches) and the many volunteers he recruited in the suburbs of Chicago, at a place called Vestal Grove. Other restoration efforts are reviewed as well: in the Everglades, the Meadowlands in New Jersey, Martha's Vineyard, and Kopta Slough on California's Sacramento River. In each project, the author highlights not only how the environment is being helped to recover but also how the volunteers themselves experience spiritual renewal. The book includes appendices with contact persons and organizations all over the country.

Grassroots Actions in Response to the Biodiversity Crisis

Clay, Jason W. **Generating Income and Conserving Resources: 20 Lessons from the Field.** Baltimore, MD: World Wildlife Fund, 1996. 76 pages. ISBN 0-89164-147-5

Simple in format, this guide contains 20 concise lessons, each revealing a simple step in the chain to reduce waste of resources during production. The book is oriented toward local communities or organizations that need to generate income but at the same time want to respect the environment and work toward social equity. The first steps include taking an inventory of community resources and making products that are already being produced and have markets. They continue with improving harvesting techniques, adding value locally (that is, processing the product on-site), using appropriate technology, and planning to make a profit, not a "killing." This is a very practical book for those interested in sustainable development.

Graham, Kevin, and Gary Chandler. **Environmental Heroes: Success Stories of People at Work for the Earth.** Boulder, CO: Pruett Publishing, 1996. 244 pages. ISBN 0-87108-866-5

Written by the owners of Earth News, an environmental news service whose stories appear in newspapers worldwide, this book profiles industries, organizations, and individuals who have integrated a commitment to the environment with their livelihood. Each chapter examines a particular activity (gardening, plastic recycling, wind power, etc.) and its proponents, and gives contact information for follow-up research.

Harker, Donald F., and Elizabeth Ungar Natter. **Where We Live: A Citizen's Guide to Conducting a Community Environmental Inventory.** Washington, DC: Island Press, 1995. 319 pages. ISBN 1-55963-377-8

This is very practical guidebook for citizen activists who need to get to the bottom of environmental contamination problems in their own communities. It is written for lay people with little background knowledge about these issues and includes a reader-friendly introduction to human impacts on the environment, organizing a citizen council, and how to begin work. The guide leads activists through the mapping process, making inventories of environmental concerns and natural resources, approaching polluting facilities, analyzing pollutants, and working with government regulatory agencies on the cleanup. The book is packed with model survey forms and worksheets, helpful diagrams, contact lists, and information tables. It will facilitate the difficult task facing anyone concerned about local contamination threats.

Johnson, Huey D. **Green Plans: Greenprint for Sustainability.** Lincoln: University of Nebraska, 1995. 207 pages. ISBN 0-8032-2579-2

Green Plans, as pioneered by the Resources Renewal Institute (see Organizations) and described by the video *Green Plans* (see Nonprint Resources), are revolutionary because they bring together many forces that taditionally have conflicts: government, business, labor, and environmentalists. Each of these groups is listened to and treated fairly in the Green Plan process. Designed as sustainable development plans with current and future generations in mind, they have met with great success where implemented. This book defines the problems Green Plans seek to solve, assesses implemented Green Plans in the Netherlands, New Zealand, and Canada, and identifies what they need to be successful. The final chapter outlines a greenprint for the United States.

Teitel, Martin. **Rainforest in Your Kitchen.** Washington, DC: Island Press, 1992. 112 pages. ISBN 1-55963-153-8

Everyone eats, and food purchases make up a large portion of most people's budget. Hence, we can help stave off the biodiversity crisis by using our consumer power: we can buy foods whose production does not lead to a decline of biodiversity. The premise behind this guide is that consumers can influence producers and middlemen. Teitel's suggestions include: (1) Buy brown eggs. Ninety percent of the eggs offered in the United States are laid by white leghorns, who lay white eggs. Buying brown eggs ensures that other breeds remain economically significant. (2) Don't insist on perfect-looking vegetables and fruits, which have been bred for their appearance but often require more pesticides, fertilizers, and water than other less beautiful but tastier, healthier, and more adaptable varieties. (3) Include more legumes in your diet, such as lentils, peas, and beans. Leguminous plants have the unique ability to return nutrients to the soil by fixing nitrogen. None of Teitel's suggestions require radical lifestyle changes—just a little more attention during shopping trips and meal planning. Why not?

Wallace, Aubrey. **Green Means: Living Gently on the Planet.** San Francisco, CA: KQED, 1994. 256 pages. ISBN 0-912333-30-8

This book includes 22 thumbnail sketches of people and projects devoted to conserving some aspect of biodiversity. The subjects range from Shaman Pharmaceuticals (see Organizations) to Steve Packard (see Biographical Sketches) to Billy Frank, a Native American man who fights to save salmon habitat with Indian fishing rights, to Catherine Sneed, who has pioneered organic gardening as therapeutic rehabilitation for prisoners. These sketches and 20 more not included in the book make up the Green Means series of five-minute public television segments, available on videotape from Environmental Media (see Nonprint Resources).

Nonprint Resources

Huge amounts of information are currently being generated in the multidisciplinary fields relating to biodiversity, and much of this information is available in a nonprint format. Specialized nonprint information sources include on-line or CD-ROM-formatted databases; computer networks and Internet resources; CD-ROMs; videotapes, videodiscs, and films; and audiotapes. This chapter devotes sections to each of these types of media and concludes with a list of distributors and service providers.

Databases

Databases, published either on-line or on CD-ROMs, are like indexes for information sources. Databases are usually searched by topic; the database will provide a list of articles or books with information addressing that topic. Researchers themselves sometimes subscribe to databases, but because subscriptions are quite costly, most people access them at libraries.

The data access industry is rapidly growing and changing, and so your local library's reference department is probably the best source for information about the databases you should consult. Most libraries subscribe either directly to databases, or to a system that comprises several databases. Many libraries are equipped to work with patrons who wish to access these databases from a home computer with a modem.

Many libraries subscribe to the Dialog service of Knight Ridder Information, Inc., a system that provides access to a variety of databases. Their Source One/UnCover® service allows users to order documents that any member library has in its collections. Information Access Company (InfoTrac) and the Online Computer Library Center (OCLC) also sell databases to libraries. Databases mentioned below that are not included in the Dialog, InfoTrac, or OCLC systems are sold by database distributors or directly by their publishers.

The following are among the databases of use for biodiversity research. Addresses for sources will be found at the end of this chapter.

Aquatic Sciences and Fisheries Abstracts

Covering all aspects of freshwater and ocean environments, this database is available through Knight Ridder Information.

BIOSIS—Biological Abstracts

Articles, books, reviews, and meeting proceedings about biology-related issues are included in this index. Available through Knight Ridder Information or directly from BIOSIS.

EBSCO MAS ELITE

MAS stands for Magazine Article Summaries; MAS ELITE provides a searchable index that yields bibliographic information, abstracts, and the full texts of some articles. Available from EBSCO.

Enviroline/Environment Abstracts

Equivalent to each other in content, Enviroline is an on-line service and Environment Abstracts comes on a CD-ROM. They include abstracts from articles on technical, scientific, socioeconomic, and policy aspects of the environment. Available through Knight Ridder Information or Congressional Information Service, Inc.

Environment

This multidisciplinary index provides bibliographic information and abstracts of articles about environmental topics. It is available through Cambridge Scientific Abstracts.

Environmental Periodicals Bibliography

Providing reference information for articles on air, land, water, and energy issues, this bibliography covers an international selection of scientific, technical, and popular periodicals. Available through Knight Ridder Information, National Information Service Corporation, and EBSCO.

Infotrac Magazine Index

Indexing a broad selection of magazines, this database provides bibliographic references and in many cases entire texts of articles. Available through Information Access Company.

Infotrac Newspaper

This citation index gives bibliographic information and a brief abstract for newspaper articles on a wide variety of topics. Available through Information Access Company.

Life Sciences Collection

Cambridge Scientific Abstracts maintains this collection of life science article abstracts. Also available through Knight Ridder Information.

National Information Center for Educational Media (NICEM)

NICEM is an international database of information about educational nonprint materials for all age levels and subject areas. Available directly from NICEM or through Knight Ridder Information.

National Technical Information Service (NTIS)

NTIS provides bibliographic information and abstracts of reports generated by government agencies such as the Environmental Protection Agency, the U.S. Fish and Wildlife Service, the Government Accounting Office, and the Department of Energy. Available through Knight Ridder Information.

NetFirst

NetFirst reviews Internet sites and indexes them by topic. Searching this database before logging on to the Internet will save time and make Internet research more effective. Available through the Epic System of OCLC.

Newsbank Science Source Collection

Directed toward high school–level researchers, this collection includes summaries of articles from 170 academic science journals. Available directly from Newsbank Reference Service.

SciSearch

A wide range of science and technology issues are covered in SciSearch's multidisciplinary index of articles published nationally and abroad. Available through Knight Ridder Information.

Social Issues Resources Series (SIRS)

SIRS provides access to articles that take clear pro or con stances on controversial issues. Available directly from SIRS.

Wildlife Worldwide

National Information Service Corporation's on-line database is searchable by geographic region and provides bibliographic information and abstracts.

Zoological Record

An index of in-print publications on animals. Available through Knight Ridder Information.

Computer Networks and Internet Resources

One of the most convenient ways to access biodiversity information is simply to "surf the web," searching with terms or words that describe your particular focus. Most government and nongovernment organizations working on biodiversity issues have set up informative homepages with hyperlinks to reports, documents, maps, and other materials of use to researchers. Electronic addresses (called URLs) for the organizations listed in chapter five of this book are included in each organization's listing. Other institutions, such as publishing houses, university environmental sciences departments, and government biological surveys, have web sites as well.

One Internet resource worth a specific mention is the Envirolink homepage (http://envirolink.org/envirohome.html), which provides links to more than 50 environmental news sources. The Environmental News Service, for example, posts six or seven new stories daily that address environmental topics—

often biodiversity-related. To subscribe to this free news service, send the following message to listproc@envirolink.org from the e-mail address where you want the articles sent: subscribe environews [Your First Name] [Your Last Name].

Subscribers to the Institute for Global Communications' e-mail and Internet access service can use a variety of networks, including EcoNet, which provides up-to-date environmental information in the form of articles, reports, conferences, and other formats. Individuals can sign up on-line or by phone.

CD-ROMs

Cities under the Sea
Date: 1995
Price: $49.95
Source: The Video Project, Environmental Media

Hosted by Jean-Michel Cousteau, this program presents the ecology of coral reefs through video, slide shows, animation, written text, and narrative. The user travels to different learning stations ("Living Communities," "Chance and Evolution," "Biodiversity," "Cycles," and "Adaptations") with an interface designed to look like a biologist's high-tech submarine.

Earth Explorer
Date: 1994
Price: $49.95
Source: The Video Project

Twenty-one environmental topics such as biodiversity, global warming, and human impact on the environment are treated in this CD-ROM. The program approaches each topic in four ways: "Explore," an interactive program, allows the user to experiment with natural world models; "Hot Topics" features debates on issues; "Articles" contains more than 400 original articles supplemented by graphics, photos, and videos; and "Data Sets" includes graphs, maps, and displays.

Earth's Endangered Environments: NGS Picture Show
Date: 1994
Price: About $40
Source: National Geographic Society, Glencoe

This CD-ROM is a bilingual (Spanish/English) introduction to the ecology of two of the world's richest ecosystems, wetlands and rain forests, and the serious threats by humankind to their survival. The information is presented in several formats: a narrated movie, a series of still photos with or without captions, and the movie's script. Classroom activities and sample tests are included; there is also supplemental background information and a glossary of terms.

Encyclopedia of U.S. Endangered Species
Date: 1995
Price: $50
Source: SVE & Churchill Media, The Video Project

This is a reference guide to more than 700 endangered species of plants and animals that are legally protected by the U.S. Endangered Species Act. Multimedia reports use written and spoken words, animal sounds, still photos, and video clips to describe each species, its habitat, and why it is endangered. Locator maps allow the user to access species by state. Study questions, quizzes, a bibliography, and a large glossary complete this useful program.

Environmental Views Series
Date: 1995
Price: $49.95 each
Source: Optilearn

The biodiversity-oriented topics covered by this series include endangered species, ocean pollution, soil erosion, and wetlands.

Eyewitness Encyclopedia of Nature
Date: 1996
Price: $75
Source: SVE & Churchill Media

Using learning games designed for students from upper-elementary to secondary levels, this multimedia encyclopedia introduces more than 250 plant and animal species and their habitats.

The Rainforest: Zooguides
Date: 1994
Price: $49.95
Source: The Video Project

The Rainforest: Zooguides describes rain forests worldwide: their ecology, the plants and animals that live there, and their human inhabitants. Also covered is rain forest destruction and why rain forests must be conserved. Photographs, video clips, maps, and quizzes are included in this CD-ROM.

Whales and Dolphins: Zooguides
Date: 1994
Price: $49.95
Source: The Video Project

This disc in the *Zooguides* series uses narration, video, animation, still photographs, and maps to describe the 70 known species of cetaceans. Includes units on life cycle, ecology, body plan, and species classification.

World Habitats
Date: 1997
Price: $149
Source: Films for the Humanities and Sciences

Habitats such as the mountains, alpine regions, Mediterranean climates, deserts, tropical mountains, the savanna, tropical and temperate forests, and polar regions are described in this disc. For each region there is a discussion of the variety of organisms living there, their adaptations and survival mechanisms, and the food webs they form. Available only for a Windows format.

Videotapes, Videodiscs, and Films

Unless noted otherwise, the media listed below are videotapes. To search specifically for 16mm films or videodiscs, consult this book's index.

General Coverage of Biodiversity

Biodiversity
Date: 1995
Length: 23 minutes
Price: $225 (purchase); $75 (rental)
Source: New Dimension Media

A British perspective on biodiversity, this program visits two sites in Britain and one in Belize. One segment shows a lab experiment in which diverse ecosystems consumed more carbon dioxide and produced more biomass than simple ecosystems. Another demonstrates "World Map" biodiversity-monitoring software in place at the London Natural History Museum. The video concludes with a review of two biological inventory projects in Belize.

Biodiversity: The Variety of Life
Date: 1992
Length: 42 minutes
Price: $150 (purchase); $40 (rental)
Source: Bullfrog Films

This introduction to biodiversity zeroes in on the Greater Northern Cascades ecosystem of Washington State and British Columbia. Concepts important to biodiversity, such as habitat fragmentation, habitat linkages, and viable populations, are defined and a variety of media is utilized, including maps and diagrams. The program's emphasis is on the larger conservation and management questions that must be solved to reach a reasonable balance between nature and development.

The Green Quiz
Date: 1992
Length: 46 minutes
Price: $295 (purchase); $65 (rental)
Source: Filmmakers Library

The Green Quiz is a test of viewers' knowledge about current threats to biodiversity and how to conserve it. Topics include global warming, deforestation, overpopulation, composting, and toxic chemicals.

Hidden Worlds
Date: 1996
Length: Series of 10 programs, 30 minutes each
Price: $29.95 each or $250 for the series
Source: PBS Video

Each program in this series spotlights a little-known animal species or survival strategy, or a unique habitat. One program focuses on the use of stealth, mimicry, and trickery by animals like cuckoos and alligators. Species-focused episodes are about poison dart frogs, penguins, antelope, and the red monkey of Zanzibar. Others visit Indonesia's Lesser Sundas and Arizona's Painted Desert.

The Infinite Voyage
Date: 1994
Length: Series of 20 programs, 60 minutes each
Price: $49.95 each
Source: Glencoe
Note: The series is also available on 10 videodiscs ($199.98 each or $1,599.98 for the series)

This secondary-level science series includes several titles pertinent to the study of biodiversity: "Life in the Balance: The Interdependence of Species and Ecosystems," visiting ancient extinction sites and current extinction hotspots; "Insects: The Ruling Class," with magnified shots of the most prevalent form of animal life on earth; "The Keepers of Eden: Preserving Endangered Species," focusing on captive breeding and reintroduction projects; and "To the Edge of the Earth: Exploring Biological Diversity," which visits five scientists in the field who study human impact on other species.

Man and the Biosphere
Date: 1990
Length: Series of 12 programs, 24 to 28 minutes each
Price: $89.95 each or $995 for the series
Source: Films for the Humanities and Sciences

This series is based on UNESCO's Man and the Biosphere Program, which examines human impact on nature and was a precursor to the sustainable development movement. Each episode focuses on a particular ecosystem (deserts, high-altitude areas, tropical rain forests, coastlines, coral reefs, wetlands, and cities), describing its features and how humans impact it.

National Audubon Video Series
Date: 1992–1995
Length: Series of five programs, 60 minutes each
Price: $28.98 each
Source: Cambridge Science

National Audubon's made-for-public-television video series includes videos about dolphin communication skills, beach pollution, caribou of the Arctic Refuge and the threat of oil and gas drilling there, the role of wildfire, and reintroduction of the wolf in North America.

Only One Earth

Date: 1987
Length: Series of 11 programs of varying lengths, grouped on five tapes
Price: $79 each (58 minutes) or $99 each (112 minutes); $249 for the series
Source: Public Media Education

This BBC series was produced in collaboration with UNESCO and originally aired on public television in the United States. Its scope is broad, describing the world's environmental problems (such as rampant fertility in Kenya and huge hydroelectric projects in Brazil) and solutions (including family carp ponds in Indonesia and preservation of wild bovines for their genetic promise in Kampuchea). Small-scale success stories as well as grand failures with disastrous and far-reaching effects are covered.

Scientific American's **Frontiers in the Environment**

Date: 1993
Length: Series of 4 programs, 20 minutes each
Price: $79 each
Source: SVE & Churchill Media
Note: The series is also available on two videodiscs for $175 each

Three of the four titles in this series on human interactions with the environment deal with biodiversity issues. "Changing the Environment" documents attempts by scientists to reintroduce the black-footed ferret into the wild and describes how ecologists construct miniature model ecosystems to facilitate ecological study. "Rescuing the Environment" shows how an Israeli marine biologist has succeeded in constructing an artificial coral reef, and how scientists helped the red-cockaded woodpecker in South Carolina survive after a hurricane knocked over most of its forest habitat. "Studying the Environment" examines the adaptations of deep sea animals and human scuba divers to the underwater environment and why sea turtles always return to the same beach to nest.

STV: Biodiversity

Date: 1996
Length: N/A
Price: $234 (Level 1) or $338 (Level III)
Source: Glencoe
Note: This program available only as a videodisc

This classroom-oriented introduction to biodiversity focuses on successes in restoration of endangered species and protection of habitat. Level I can be used with just a laser-disc player; Level III is more interactive and requires a computer hookup.

Web of Life
Date: 1995
Length: 113 minutes
Price: $46.95
Source: World Wildlife Fund

The variety of approaches to studying and protecting biodiversity is illustrated in this tapestry of over a dozen short portraits of biodiversity-oriented people and projects. A desert research project has pinpointed kangaroo rats as the keystone species of that habitat. Wildflower experiments in Minnesota have proved that diverse communities are more stable than simple ones. A photographer documents the conversion of *Homo sapiens* into "Techno sapiens" and expresses concern about their resulting passivity in the face of disappearing biodiversity. A mother-son team of ecotourists are forever changed by their voyage to the Amazon. Some of the best-known faces in biodiversity, including E. O. Wilson, Terry Erwin, and Thomas Lovejoy, make cameo appearances. Accompanied by an informative secondary teacher's guide and glossy student primer, this video serves as a lively, effective introduction to the topic.

Evolution of Species and Their Adaptations

The Burgess Shale
Date: 1992
Length: 45 minutes
Price: $270 (purchase); $60 (rental)
Source: New Dimension Media

The Burgess Shale site in the Canadian Rockies is rich with animal fossils from the early to middle Cambrian period, more than 500 million years ago. The differences between body forms of that time and the present have led scientists to wonder whether the phyla of that period were vastly different than those existing now. This video introduces evolutionary theory, focusing particularly on the evidence at the Burgess Shale site.

GAIA: The Living Planet
Date: 1989
Length: 54 minutes
Price: $149 (purchase); $75 (rental)
Source: Films for the Humanities and Sciences

This video introduces the Gaia Hypothesis, which postulates that the earth's chemical elements and biological organisms work together as an active regulating mechanism. The precise mixture of carbon dioxide, nitrogen, oxygen, and methane in the atmosphere results in conditions perfect for life. Plants and animals fluctuate their levels of emission and absorption of the elements to maintain an ideal atmospheric temperature. The father of the Gaia Hypothesis, Dr. James Lovelock, is interviewed at his home, and there is a visit to the Marine Biological Laboratory in Plymouth, England, where research confirms that when atmospheric temperatures increase, plankton increase in numbers and in collective intake of carbon dioxide. This results in greater output of oxygen and a cooling of the climate.

Life Is Impossible
Date: 1993
Length: 50 minutes
Price: $34.95
Source: BBC

What were the ingredients and conditions that jump-started life on earth? The video starts with the "primordial soup" experiment in which scientist Stanley Miller subjected a mixture of inorganic compounds to an electrical charge and simple life-forms bloomed. Other scientists respond to the theory that developed out of this experiment, introducing more evidence and alternative hypotheses.

Lifesense
Date: 1991
Length: Series of five programs, 30 minutes each
Price: $19.95 each or $98.95 for the series
Source: BBC

Using creative photographic methods and computer-generated special effects, this series reveals worlds experienced by different animals, including flies, vultures, mites, honeybees, otters, and macaque monkeys. Each episode makes a specific journey: "Home Life" shows how animals adapt and depend upon one another for food and shelter; "Seeds of Life" demonstrates how plants use animals to disperse their seeds and why animals willingly cooperate in the process; "Partners for Life" explores the relationship between humans and domesticated animals; "Human Life" suggests that there are more organisms living on the human body than people on earth; and "Life in the Balance" demonstrates how humans disrupt habitats and food webs.

Ourselves and Other Animals

Date: 1995
Length: Series of 12 programs, 27 minutes each
Price: $89.95 each or $999 for the series
Source: Films for the Humanities and Sciences

Filmed at wilderness sites all over the world, this series spotlights a range of animal behaviors including courtship, "high-tech" hunting techniques, aggression, communication, deception, and territoriality, all of which are adaptations for survival.

Survival in Nature

Date: 1990
Length: Series of 5 programs, 26 minutes each
Price: $275 each or $1,195 for the series (purchase); $50 each or $160 for the series (rental)
Source: New Dimension Media

Each of the videos in this series explores the adaptations of plants and animals to their environments. "Life on the Edge" visits Alaskan mountain goats that live only at very high altitudes and depend on food sources no other animal uses. "The Silver Trumpeter" profiles the trumpeter swan, a species that barely escaped extinction in the early part of this century. After their 1918 low point, a refuge was set aside as a breeding ground and the trumpeters managed to build up a viable population. "Technical Animals" documents the small number of species, which, along with humans, have learned to use tools. "Venom" distinguishes venom from poison and shows how animals ranging from poison dart frogs to salamanders to scorpions have developed chemical warfare. "When the Tide Goes Out" features tide pool creatures of the intertidal zone: barnacles, crabs, and starfish. These unique organisms survive because they have successfully adapted to an environment with drastically variable cycles.

Survival of the Fittest

Date: 1990
Length: 24 minutes
Price: $89.95
Source: Films for the Humanities and Sciences

Aggressive behavior does not always lead to the demise of losers; sometimes individuals resolve conflicts or defuse them quickly in order to avoid fighting. This program examines aggressiveness and passivity as survival strategies.

Descriptions of Biodiversity in Different Ecosystems

Alaska's Rain Forest: The Tongass
Date: 1993
Length: 26 minutes
Price: $270 (purchase); $60 (rental)
Source: New Dimension Media

The subject of this program is the largest temperate rain forest in the world, which extends throughout southeastern Alaska. The plant, animal, and human ecology of this area is described, with an emphasis on the complex interactions between them.

America's Rainforest
Date: 1992
Length: 20 minutes
Price: $49.95
Source: Cambridge Science

The continental United States has only one temperate old-growth rain forest, located on the Olympic Peninsula of Washington State. This video presents its fragile ecology and diverse plant and animal inhabitants, along with the threats to the forest by the commercial logging industry.

Coral Reefs: Rainforests of the Sea
Date: 1996
Length: 20 minutes
Price: $59.95 (purchase); $35 (rental); low-income discount
Source: The Video Project

This program introduces the ecology of coral reefs and their importance to marine biodiversity. It shows where reefs are found, how they form, and how human activity threatens them. This version was created for secondary school and adult audiences. Another, entitled *The Amazing Coral Reef* (same length, price, and source), was designed for elementary school students.

Coral Reefs: Their Evolution and Reproduction
Date: 1994
Length: 27 minutes
Price: $275 (purchase); $60 (rental)
Source: New Dimension Media

Coral reefs are the most colorful ecosystems on the planet, possibly harboring as many species as tropical rain forests. This video focuses on coral's little-known reproductive process, which is instantaneous and takes place just once a year. The evolutionary strategy that has allowed it to survive and flourish on shallow ocean floors is featured as well.

Desert under Siege
Date: 1991
Length: 28 minutes
Price: $79 (purchase); $40 (rental); low-income discount
Source: The Video Project

Challenging the stereotype of deserts as barren wastelands, this video introduces desert ecology, gives a history of human inhabitation of deserts, and examines the environmental impact of such human activities as urbanization, strip-mining, military training, livestock grazing, and off-road vehicle recreation.

Ecosystems in Action
Date: 1995–1997
Length: Series of 4 programs, 21 to 24 minutes each
Price: $79 each or $237 for the series
Source: Hawkhill Video

The four on-site ecosystem descriptions include a tropical rain forest in Amazonia, a New England pond, the Galápagos Islands, and the Great Lakes. The Galápagos video includes a step-by-step history of how life colonizes barren islands: first amphibious plants creep from the sea to rocky outcrops; eventually they create primitive soil; this allows terrestrial plants to take root; and plants in turn attract consumer species like birds and lizards. The Great Lakes video charts the effects on lake ecology of dense human population and exploitation of the area's rich mineral resources. Each video comes with a detailed viewer's guide.

Finite Oceans
Date: 1995
Length: 60 minutes
Price: $24.95
Source: Cambridge Science

If we were to reenvision the oceans as finite bodies of water, we might stop using them indiscriminately as toxic waste dumps. This Discovery Channel video shows how even minute quantities of DDT, PCBs, and pesticides in the ocean travel up the food chain and threaten all forms of life.

Forests for the Future
Date: 1995
Length: Series of three programs, 21 to 27 minutes each
Price: $59.95 each or $149 for the series (purchase); $30 each or $75 for the series (rental); low-income discount
Source: The Video Project

The first video in the series, "The Natural Forest," visits an old-growth forest and shows how it shelters an interconnected web of species, from microorganisms to huge mammals. The second, "Humans in the Forest," examines the impact of clear-cutting and reforesting on the ancient forests of the Pacific Northwest. Finally, "Decisions for the Future" scans some of the ways that humans can utilize forest resources sustainably.

The Living Planet
Date: 1994
Length: Series of 12 programs, 60 minutes each
Price: $99.95 each or $595 for the series
Source: Cambridge Science

Dynamic host David Attenborough takes viewers on a tour of the planet's major biomes, from northern boreal and tropical forests to deserts to grasslands to freshwater habitats and the ocean. Forty videographers filmed this series in 63 countries.

The Natural World of Latin America
Date: 1995
Length: Series of 11 programs, 26 to 28 minutes each
Price: $99 each or $995 for the series
Source: Films for the Humanities and Sciences

Each program visits a different region of South, Central, or North America, describing its animal and plant inhabitants and the adaptations they have evolved. Sites include the Sea of Cortez, the Sonoran Desert, Amazonia, the Pampas and the Chaco, the Andes, South American wetlands, and Antarctica.

Our Fragile World
Date: 1997
Length: Series of four films, 15 minutes each
Price: $69.95 each or $249 for the series
Source: Films for the Humanities and Sciences

The videos in this series each hone in on the inner workings of a specific ecosystem and the challenges it faces from human impact. Ecosystems included are the Khutzeymateen Valley south of the

Alaska panhandle, Valdez Bay along Argentina's central Atlantic Coast, Clayoquot Sound off Vancouver Island, and Costa Rica's rain forests.

Riddle in the Sand
Date: 1995
Length: 50 minutes
Price: $34.95
Source: BBC

Set at a Scottish estuary, *Riddle in the Sand* traces the flow of energy from the sun through an intricate food web in which all organisms are interdependent. The questions of how humans interrupt natural cycles, and when they should intervene, are also addressed.

Secrets of the Bay
Date: 1990
Length: 28 minutes
Price: $59.95; low-income discount
Source: The Video Project

The fragile San Francisco Bay ecosystem is home to 6 million humans and thousands of other species. This video highlights the impact of urbanization on the bay but also includes footage of several spectacular animal species, including a peregrine falcon nesting on the Bay Bridge, harbor seals who live in the Bay's marshland, pelicans, and avocets.

Secrets of the Choco
Date: 1995
Length: 52 minutes
Price: $250
Source: Bullfrog Films

The Choco rain forest of Colombia is one of the largest and most isolated ecosystems of this type in the world. Six tropical biologists accompany the cameraman, explaining the forest's intricate ecology and debating its future.

Secrets of the Rain Forest: The Tree Canopy
Date: 1997
Length: 50 minutes
Price: $149 (purchase); $75 (rental)
Source: Films for the Humanities and Sciences

Tropical rain forest soil is nutrient-poor except for a thin humus layer on top of the ground. The richness of the rain forest is found in its canopy, where myriad plant and animals coexist and interact.

Secrets of the Salt Marsh
Date: 1994
Length: 20 minutes
Price: $150 (purchase); $20 (rental)
Source: Bullfrog Films

Salt marshes protect the mainland from floods and filter sediment and pollutants from the water. Many migratory species use them as rest stops, and they provide a permanent home for numerous other aquatic and avian species. This program demonstrates the interrelations between species that inhabit salt marshes and the importance of salt marshes to other ecosystems.

Treasures of the Greenbelt
Date: 1987
Length: 28 minutes
Price: $59.95 for institutions; low-income discount
Source: The Video Project

Greenbelts surround many American communities, and San Francisco is fortunate to be bounded by 4 million acres of open space in the form of parks, farms, ranches, and vineyards. This video is an exploration of the human and ecological diversity of San Francisco's greenbelt, explaining its importance both in environmental terms and in terms of how it improves quality of life.

Tropics in Trouble
Date: 1990
Length: 25 minutes
Price: $12.98
Source: Turner MultiMedia

Four segments describe some of the tropical ecosystems of Costa Rica and attempts to preserve them. A biological prospector troops through La Selva, in the northern zone of the country, in search of plants that may serve as natural pesticides. Scientists track the migratory patterns of the resplendent quetzal, symbol of freedom for Central Americans and trophy species for ecotourists. Only its mating ground in the high-altitude cloud forests was previously known, but after biologists found out that it migrated to lowland forests, they sought to preserve those ecosystems as well.

Another segment describes the ambitious restoration plan for the dry tropical forest spearheaded by biologist Daniel Janzen. (See profile in Biographical Sketches and review of *The Quetzal and the Macaw* in Print Resources.)

The Unique Continent: Australian Environmental Studies
Date: 1992
Length: Series of 13 programs, 30 minutes each
Price: $95 each or $1,075 for the series
Source: Landmark Media

These programs, viewed together, provide a complete picture of Australia's environment: its ancient past, its present, and its future. Individual episodes focus on such issues as the importance of the smallest-sized species (such as termites, ants) to the overall health of ecosystems, aboriginal environmental practices, plant and animal invaders from other continents, and environmental controversies.

The Wasting of a Wetland
Date: 1991
Length: 23 minutes
Price: $195 (purchase); $40 (rental)
Source: Bullfrog Films

This is a description of the plunder of the Everglades—habitat for uncountable species, waystation for migratory birds, and the only source of fresh water for south Florida—where the human population is increasing faster than anywhere else in the United States. Consumer advocate Ralph Nader calls this production "an informative and haunting documentary that exposes legislative loopholes and the short-sighted self-interest of the agricultural and development industries."

Wetlands and Natural Resources
Date: 1995
Length: 47 minutes
Price: $29.95
Source: Turner MultiMedia

Fourteen CNN reports on wetlands visit different sites and report on them from unique angles. Among the topics are tours of the Ofekenokee and Honey Island swamps in the Deep South; why amphibian populations are currently declining so rapidly; how restorationists in Tennessee and outside San Francisco are re-creating wetlands; threats to the best-known wetland in the United States, the Florida Everglades; and federal policies regarding wetlands.

Focus on Certain Taxa

The Bug Man of Ithaca
Date: 1991
Length: 45 minutes
Price: $195 (purchase); $50 (rental)
Source: New Dimension Media

"The Bug Man" is Dr. Thomas Eisner, renowned entomologist and expert on insects' use of chemicals for defense and communication (see Biological Sketches). Host David Suzuki follows Eisner around his Cornell University lab and central Florida research site as Eisner, with childlike delight, reveals secrets of the insect world. For example, flowers and bugs show one pattern to us but sport a completely different one under ultraviolet light—a spectrum that insects perceive but we do not. These UV patterns help insects find nectar and each other for mating. Some caterpillars actually sew themselves a costume from flower petals to look like a part of the flower they are devouring. Zoom videography shows carnivorous plants in action and spiders attacking insects that Eisner flicks one after another into their webs. Interspersed through the segments on insect mimicry, the amazing adaptations they've evolved, their chemical defenses, and the lack of funds for this type of research, are insect horror movie clips. By the end of the video it is clear that insects are fascinating, complex creatures vital for the survival of the earth and that Hollywood's horrifying version of them is ill-informed and vulgar.

Fungi: Friend and Foe
Date: 1992
Length: 25 minutes
Price: $280 (purchase); $50 (rental)
Source: New Dimension Media

The wide variety of fungi is examined here, both friend (such as the fungi that wrap themselves around plants' roots and provide them with necessary chemical nutrients) and foe (including the damp rot that is eating away at the medieval murals of Canterbury Cathedral). Certain fungi attack crop pests like taro beetles or locusts, and scientists are researching ways to introduce those species of fungi in places where crops suffer those plagues. The life cycle of fungi is detailed, as are the experimental trials in laboratories to discover how much more useful fungi could become in agriculture and medicine.

Insects
Date: 1990
Length: 23 minutes
Price: $89.95
Source: Films for the Humanities and Sciences

This program first examines social insects like bees, whose pollinating and honey-making activities are of obvious help to humans. It then turns to mosquitoes and black flies, which do not garner as much appreciation from people but do occupy important positions in their ecosystems.

The Life of Plants
Date: 1988
Length: Series of 13 programs, 28 minutes each
Price: $89.95 each or $1,095 for the series
Source: Films for the Humanities and Sciences

The Life of Plants is an exquisitely photographed series showing plants' diversity and intricate adaptations. Episodes cover a range of plant-oriented topics, from "The Life Cycle of Plants" to "Adaptation to Site" to "Plant Defenses" to "The Balance of Nature."

Plankton: The Breathing Sea
Date: 1986
Length: 26 minutes
Price: $280 (purchase); $50 (rental)
Source: New Dimension Media

Even more important than rain forests in maintaining the right balance of gases in the atmosphere are plankton. These tiny organisms, abundant in every ocean, consume carbon dioxide and excrete oxygen. Years ago, single-celled plants between one and three millimeters long, with names like diatoms and dinoflagellates, were thought to be the smallest forms of phytoplankton. Now, thanks to today's more powerful microscopes, we know that there are other plants 1 million times smaller, and that these organisms may make up half of all the plant material in the ocean. This video provides magnified images of some of the magnificent forms of plankton, discusses the logistical difficulties in studying them, and illustrates the importance of this base of the food web for the animals that directly consume them as well as the rest of us whose survival depends on an atmosphere made up of a particular balance between oxygen and carbon dioxide.

Predator

Date: 1994
Length: 30 minutes
Price: $34.95
Source: BBC

The Polynesian archipelago has been taken over by the Euglandina African land snail—one of the fiercest and most destructive snail species. It has seriously threatened all of the delicate native *Partula* species and destroyed some of them. This documentary gives a graphic example of what can go wrong when nonnative species are introduced—for any reason. See also Stephen Jay Gould's book *Eight Little Piggies* (discussed in Print Resources) for a description of this problem.

The Private Life of Plants

Date: 1995
Length: Series of six programs, 60 minutes each
Price: $89.95 for the series
Source: Turner MultiMedia

A worldwide exploration of the plant kingdom, these programs use the latest technology in film, video, and computers to show plants from a variety of perspectives: from altitudes of hundreds of feet above them, under powerful microscopes, and from what animators imagine to be plants' own perspective. This series, hosted by David Attenborough, was previously broadcast on television by TBS.

Spirit of Trees

Date: 1992
Length: Series of eight programs, 25 minutes each
Price: $195 each or $995 for the series (purchase); $50 each or $195 for the series (rental)
Source: The Cinema Guild

This series, produced in Great Britain, examines trees and what they do for other forms of life. Engaging host Dick Warner interviews conservationists, scientists, folklorists, woodsmen, seed collectors, forest rangers, wood turners, charcoal makers, and foresters. His exploration reveals that the people of the British Isles hold a deep respect for trees, based on a long history of considering them valuable natural resources.

Threats to Biodiversity

The Amazon: A Vanishing Rainforest
Date: 1988
Length: 29 minutes
Price: $250 (purchase); $50 (rental)
Source: The Cinema Guild

The camera follows scientists from the National Institute for Amazonian Research, a Brazilian laboratory charged with studying the Amazon Basin's ecology and proposing development strategies. One of their discoveries deals with erosion of topsoil. With the forest cover intact, there is very little erosion or water runoff. But when the trees are cut down, ten times more water runs off the land, and the runoff contains ten times as much sediment. Unfortunately, this finding has not deterred government planners, who envision hundreds of dams producing enough electricity to supply Brazil and its neighbors. However, eroded sediment fills up the dams' reservoirs quickly, and the water, with a high content of sulfuric acid absorbed from rotting trees covered by the reservoirs, makes the dams functional for fewer than ten years. This is a brief overview of the environment-versus-economics quandary in which many tropical countries find themselves.

Burning Rivers
Date: 1992
Length: 28 minutes
Price: $85 (purchase); $40 (rental); low-income discount
Source: The Video Project

Guatemala's serious environmental problems (including deforestation, river pollution, and poisoning of farm workers with pesticides) are shown to result primarily from the inequity between rich and poor there.

Can Polar Bears Tread Water?
Date: 1989
Length: 58 minutes
Price: $295 (purchase); $90 (rental)
Source: The Cinema Guild

Two hundred years of industrialization have accelerated the buildup of greenhouse gases in our atmosphere, and our climate has already started to change. An increase in average global temperatures of two to five degrees centigrade could occur within the next 50 years. Increases would be more marked near the pole, and the polar ice caps could even melt. If they did, cities where three-fourths of the world's human population lives would be inundated; most coastal wetland habitats would be eliminated; and polar bears and other polar creatures would probably drown. Despite the dour outlook, scientists interviewed in this program are optimistic.

The Decade of Destruction

Date: 1990 (unabridged version); 1991 (classroom version)
Length: Available as a series of five programs of 51 to 57 minutes each (unabridged version), or as a series of six programs of 10 to 19 minutes each (classroom version)
Price: Unabridged version: $495 (purchase), $375 (rental); classroom version: $350 (purchase), $75 (rental)
Source: Bullfrog Films

The 1980s were the decade of destruction in the Amazon Basin rain forest, and filmmaker Adrian Cowell was there to record it. The unabridged version of this video has appeared on PBS's "Frontline," and the classroom version distills the details into shorter segments. Both versions document the struggle for Amazonia from the perspectives of the people involved: indigenous people trying to hold onto their age-old forest-dwelling traditions but usually succumbing to physical disease or losing their culture; desperate peasants who came to the forest with hopes for a new future—and watched it drain away once the land they cleared became barren and infertile; absentee landowners who make up 1 percent of the population yet hold a quarter of the country's arable land; politicians who believed Brazil's path out of poverty would be paved by Amazon resources; environmentalists who tried to halt the politicians' plans; miners, both the individualistic, freelance breed and those associated with multinational companies, in pursuit of the world's largest untapped reserve of gold; and rubber tappers, inventors of the "extractive reserve" concept through which rain forest resources could be harvested sustainably. Despite the harm that has already been done, the mood of this series is hopeful.

Endangered

Date: 1994
Length: 30 minutes
Price: $29.95 (purchase); $20 (rental); low-income discount
Source: The Video Project

Produced by the National Wildlife Federation, this video explains how human impact has led to the loss of 100 species per day, why this situation is so tragic, and how the Endangered Species Act has helped restore populations of more than 200 endangered species. A few well-known scientists, including E. O. Wilson (see Biographical Sketches), make short appearances.

Endangered Species

Date: 1991
Length: 30 minutes
Price: $24.98
Source: Turner MultiMedia

Endangered species are threatened for a diverse set of reasons, and this video delves into some of them. The Rwanda mountain gorillas, for example, whose diminishing habitat has made them prisoners of a very small refuge, face an additional threat from poachers in this hungry, war-torn nation. Three thousand sea lions off the coast of California are killed every year when plastic fishing line discarded by fishermen chokes them or cuts into their flesh. Colorful parrots are poached in their native forests and smuggled into south Florida. Four to five animals die for every one that becomes someone's pet. Black-footed ferrets were driven to near extinction by canine distemper spread by domestic dogs and by urbanization, which diminished the population of their primary prey, prairie dogs. Each segment is just a snapshot, but together they can introduce an in-depth study of the problems leading to endangerment.

Endangered Wildlife

Date: 1993
Length: Series of six programs, 60 minutes each
Price: $19.98 each or $112.98 for the series
Source: Cambridge Science

Produced by the National Geographic Society, each video in this series visits a well-known endangered species in its habitat: the elephant, wild pandas, great whales, mountain gorillas, and black rhinos.

Growing like a Weed
Date: 1992
Length: 29 minutes
Price: $59.95
Source: The Video Project

Some of the most bloodthirsty attacks on biodiversity occur in suburbia, where deadly pesticides are enlisted to keep weeds from growing in lawns and backyard gardens. Toxicologists, landscape architects, and people sensitive to herbicides testify against chemical-based gardening.

In a Time of Headlong Progress
Date: 1994
Length: 45 minutes
Price: $125 (purchase); $45 (rental); low-income discount
Source: The Video Project

More than 90 percent of the lush coastal rain forest ecosystem of the Bahía state of Brazil has disappeared already. The golden-headed tamarin, endemic to the Bahía area, is perhaps the cutest of endangered species and has inspired Bahíans to preserve what is left of its native habitat. Conservationist Cristina Alves has done much to make the tamarin into a symbol for Bahíans and is trying to convince them that their own well-being is tied into that of the tamarin. She and other conservationists are engineering a model for sustainable development that could lead to the restoration of lost rain forest and a healthier regional economy.

The Killing Tide
Date: 1993
Length: 40 minutes
Price: $29.95
Source: Turner Educational Services

Fish populations throughout the world are rapidly declining, and this video shows why. It takes the viewer onto a family-owned vessel out of Glouster, Massachusetts, that has so many technological aids for locating fish that entire schools can be scooped out of the ocean. Fisheries depletion has decreased the catch dramatically over the past several years, but instead of examining their own methods, the fishermen blame conservationists and the government for imposing limits on them. Another segment shows the waste on fishing boats. Tiny one-year-old red snapper die in a shrimp net or a heavily salinized sorting bin; the biologist shown sorting through the dead young grimly reports that they would have been an important catch several years later. As if fish weren't threatened enough directly, their spawning grounds are being destroyed when coastal wetlands are filled in for residential and agricultural development. The only solution to the problem seems to be more intelligent land development and voluntary self-regulation on the part of the fishing industry, but the video shows how complicated those questions are. All in all, this is a portrait of an irresponsible industry whose life span will soon be cut short.

The Last Show on Earth

Date: 1993
Length: 102 minutes
Price: $295 (purchase); $85 (rental)
Source: Bullfrog Films
Note: Also available also as a series of four videotapes, 27 minutes each; $395 (purchase), $95 (rental)

Famous personalities and musicians decry the loss of biodiversity in this star-studded feature. Three of the four segments describe the problem ("Endangered Species," "Endangered Habitat," and "Endangered Cultures"), and the last, "Regaining Balance," calls on humans to change our destructive behavior. Features appearances by Ben Kingsley, Sigourney Weaver, Chief Red Cloud, and the Dalai Lama; music is provided by Ladysmith Black Mambazo, Carlos Nakai, Peter Gabriel, Kate Bush, Vangelis, Julian Lennon, and Elton John.

Lester Brown: Assessing Our Planet's Health

Date: 1990
Length: 30 minutes
Price: $89.95
Source: Films for the Humanities and Sciences

Lester Brown, head of Worldwatch Institute (see Organizations), gives a sobering assessment of the planet's health: increased air and water pollution, more toxic waste and poisoned wildlife, fewer trees, more desert, and less arable land. Brown emphasizes that nations must cooperate in the solution to these problems and that every country should design an environmentally sustainable path to economic health.

Lucia
Date: 1994
Length: 89 minutes
Price: $150 (purchase); $75 (rental)
Source: Bullfrog Films

Lucia is a feature-length dramatization of the ties between a community and its local ecosystem. A tanker spills oil in the sea off a coastal fishing village in the Philippines, and the townspeoples' lives are completely ruined. They get sick, their animals get sick, and fishing becomes impossible. Families disintegrate and migrate to Manila's shantytowns. This is one film in the series *Developing Stories,* which gives voice to filmmakers of the developing world.

Our Threatened Heritage
Date: 1989
Length: 19 minutes
Price: $29.95
Source: The Video Project

Produced by the National Wildlife Federation, this overview of tropical rain forest destruction links flawed development efforts with environmental degradation and proposes some innovative solutions.

Protecting Our Endangered Species
Date: 1996
Length: 30 minutes
Price: $99
Source: New Castle Communications

Aimed at students in grades four through eight, this child-narrated production emphasizes the interconnectedness between the human species and all others on earth, guiding viewers to learn about wild habitats and what threatens them. It includes a 20-page teacher's resource guide. New Castle produces a similar program geared for K–3 students, "Being an Explorer" (same date, price, and length).

Richard Leakey: The Man Who Saved the Animals
Date: 1995
Length: 50 minutes
Price: $89.95
Source: Films for the Humanities and Sciences

When archaeologist Richard Leakey became the head of Kenya's national park system in the late 1980s, he was immediately challenged with the widespread and highly publicized problem of elephant poaching. This program documents his battle with the poachers and examines the underlying social, economic, and political factors that lead to wildlife depletion in Kenya.

Saving the Manatee
Date: 1994
Length: 15 minutes
Price: $19.95
Source: Turner MultiMedia

The manatee, sometimes called the sea cow, is a giant aquatic mammal that lives in shallow coastal habitats. Because of its size and calm nature, it has in the past been a prime source of meat for coastal dwellers. It is now protected in the United States by the Endangered Species Act, but careless motorboat drivers still threaten its small population, and poachers target the manatee as well. This report describes the manatee, shows how scientists track its migration routes and population changes, and visits centers where manatees are protected and studied.

Scientific American's Frontiers in Science and Nature
Date: 1990
Length: Series of four programs, about 20 minutes each
Price: $79 each
Source: SVE & Churchill Media
Note: The series is also available on two videodiscs ($175 each)

Two of the programs in this series treat biodiversity issues. "Animal Behavior" describes bizarre behavior evolved by two species: ants that kidnap unborn babies from other nests and enslave them when they hatch; and a species of squirrel that somehow manages to survive despite the fact that its body temperature dips below freezing during its yearly hibernation. "Survival in Nature" documents efforts to protect Old World migratory storks and foxes of Catalina Island off California, both of which are endangered.

Spaceship Earth
Date: 1990
Length: 25 minutes
Price: $39.95
Source: The Video Project

Produced with young viewers in mind and hosted entirely by kids themselves, *Spaceship Earth* travels the planet examining the interconnected biodiversity issues of deforestation, global warming, and ozone depletion. British rock star Sting appears briefly to advocate the protection of rain forests and their indigenous inhabitants.

Still Life for Woodpecker?

Date: 1992
Length: 28 minutes
Price: $250 (purchase); $50 (rental)
Source: Bullfrog Films
Note: Also available in 16mm film format; $550 (purchase) $50 (rental)

Weaving together Native American beliefs from the Pacific Northwest and forest ecology, this award-winning video examines the rapid loss of old-growth forests in the United States. According to Native American cosmology, the pileated woodpecker was sent by the Creator to watch over the forests. Modern forest ecologists have realized that pileated woodpeckers are both an important member of the forest's food web and an indicator species for the health of old-growth forests. The program issues a sharp critique of the U.S. government's traditional forest management policies.

Turning the Toxic Tide

Date: 1994
Length: 27 minutes
Price: $195 (purchase); $45 (rental); low-income discount
Source: Bullfrog Films

Jobs and environmental protection are often placed in opposition: jobs lead to environmental degradation and environmental protection takes away jobs. But this video shows how degradation caused by a paper mill upstream robbed clam diggers and oyster diggers working at the mouth of a river in British Columbia of their livelihood. The message to viewers is that through responsible purchases (such as recycled paper) consumers can make a difference.

Yosemite and the Fate of the Earth

Date: 1993
Length: 14 minutes
Price: $95 (purchase); $25 (rental)
Source: Bullfrog Films

Even in Yosemite National Park, global environmental problems are manifested locally. Fewer migratory songbirds grace the park due to diminishing habitat along their route southward. Campfires cause serious air pollution. One-third of the park's trees have been damaged by emissions from tourists' vehicles and nearby industry. And amphibians are disappearing here as they are throughout the rest of the world.

Indigenous Peoples and Biodiversity

Ancient Futures: Learning from Ladakh
Date: 1993
Length: 59 minutes
Price: $95 (purchase); $45 (rental); low-income discount
Source: The Video Project

For more than 1,000 years, people have lived well in Ladakh, high in the western Himalayas of India. There is no pollution or social strife, thanks to their respect for the environment and their ability to cooperate with one another. Western civilization is reaching Ladakh, however, and unless this model society can survive intact, the rest of the world may never learn its secrets to sustainability.

Blowpipes and Bulldozers
Date: 1989
Length: 60 minutes
Price: $250 (purchase); $85 (rental)
Source: Bullfrog Films
Note: Also available in 16mm film format; $850 (purchase); $85 (rental)

The nomadic Penan people of Borneo have lived in their rain forest ecosystem for what researchers estimate to be 40,000 years. Currently, however, their way of life and their habitat are threatened with rapid decimation by loggers. This film, made clandestinely, documents their unique lifestyle and the dangers the Penan now face. It features their foreign guest, Bruno Manser, a Swiss man who has lived with the Penan people for half a decade and who has been instrumental in publicizing their plight.

Cry of the Forgotten Land
Date: 1995
Length: 26 minutes
Price: $89 (purchase); $45 (rental); low-income discount
Source: The Video Project

A documentary about the Moi people of Indonesian New Guinea's rain forests, who have lived in isolation from other civilizations until very recently. The Moi, their forest, and the 100,000 endemic species found there are endangered by a repressive government and a massive logging industry.

Green Medicines

Date: 1994
Length: 52 minutes
Price: $149 (purchase); $75 (rental)
Source: Films for the Humanities and Sciences

The path of botanical medicines from their original source to pharmacy shelves is long and complex. This video traces that path, first covering visits by ethnobotanists to rain forests and their indigenous inhabitants in Samoa, Thailand, Borneo, Papua New Guinea, and Brazil. The researchers collect plants and compile information about them from indigenous shamans. The specimens then are sent to pharmaceutical laboratories in the United States and Europe, where their active compounds are isolated. After intense experimentation, the medicines are forwarded to research hospitals, where doctors use them on human patients.

In Good Hands: Culture and Agriculture in the Lacandon Rainforest

Date: 1994
Length: 27 minutes
Price: $125 (purchase); $45 (rental); low-income discount
Source: The Video Project

This program pays a visit to Lacandon farmers of Mexico's Chiapas rain forests. The Lacandon cultivate their crops in a delicate rain forest ecosystem and have been doing so, sustainably, for centuries. Unfortunately, very few individuals still use these methods. The camera follows ecological anthropologist Dr. James Nations, who talks with Lacandon elders about their farming methods and how their culture, mythology, and religion influence their approach to agriculture.

Jungle Pharmacy

Date: 1989
Length: 53 minutes
Price: $295 (purchase); $90 (rental)
Source: The Cinema Guild

The Amazon Basin is like a giant medicine chest, and the indigenous peoples living there know thousands of botanical remedies. *Jungle Pharmacy* makes several assertions pertinent to biodiversity: indigenous people are the best caretakers of their forest homes; the medical knowledge of their shamans is exploited by ethnobotanical and medical researchers to benefit people elsewhere, though at the source, forests and indigenous cultures are in great danger; indigenous shamans in South America have organized to reclaim their medical traditions. The cast of characters includes Western ethnobotanists and pharmacists and many indigenous people from Brazilian and Peruvian Amazonia, including two charismatic shamans.

The Moon's Prayer
Date: 1991
Length: 51 minutes
Price: $85 (purchase); $35 (rental); low-income discount
Source: The Video Project

From the perspective of many Native Americans, their war with white settlers for control of natural resources is not over yet. This program shows tribes of the Pacific Northwest in a struggle to regain their tribal lands. Their use of modern methods and tools to restore the health of their fishing grounds and homeland has given their white neighbors the opportunity to learn much about true sustainability.

Oren Lyons the Faithkeeper
Date: 1997
Length: 60 minutes
Price: $89.95
Source: Films for the Humanities and Sciences

Bill Moyers talks with environmentalist Oren Lyons of the Onondaga tribe, who discusses his peoples' legends, prophecies, and traditions of wisdom. Lyons speaks about the sacredness of the earth, how the Great Law of Six Nations holds that humans are part of their environment, and how Native Americans establish community.

Yanomami: Keepers of the Flame
Date: 1992
Length: 58 minutes
Price: $95 (purchase); $45 (rental); low-income discount
Source: The Video Project

The Yanomami people of the Amazon Basin live as one of the world's last "untouched" cultures. This video documents a visit to their settlement by a group of journalists, anthropologists, and doctors and gives an in-depth examination, from the point of view of the visitors, of the Yanomami way of life. The

Yanomami's lands are soon to be encroached upon by gold miners, who will bring another lifestyle as well as disease and environmental destruction. The video concludes with an appeal to protect the Yanomami and all other indigenous cultures.

Agriculture and Biodiversity

Diet for a New America
Date: 1991
Length: 30 minutes
Price: $19.95
Source: The Video Project

The average American diet not only results in widespread obesity and disease, it also encourages agricultural practices that do great damage to the environment. The host of this program is John Robbins, founder of EarthSave International (see Organizations) and articulate advocate of a diet based primarily on plant protein.

Food or Famine?
Date: 1997
Length: Series of 2 programs, 49 minutes each
Price: $495 (purchase); $95 (rental)
Source: Filmmakers Library

How to feed a growing world population while using sustainable agricultural methods is a puzzle yet unsolved. High-yield methods have led to erosion, salinization, and chemical pollution, all factors in the loss of biodiversity. In response, many farmers have returned to traditional agricultural methods that do little damage to the land but do not produce as much. This video visits farms in North America, Chile, Indonesia, Africa, and India. Commentators Lester Brown of the Worldwatch Institute (see Organizations) and others reflect on how enough food can be produced in the long term for a rapidly growing population.

Fruits of the Earth: Promising Plants for Tomorrow
Date: 1994–1996
Length: Series of 21 programs, 15 minutes each
Price: $175 each or 10 for $1,550
Source: Landmark Media

The programs in this series each feature a plant that promises to relieve hunger or help heal scarred agricultural lands: jojoba, potato, banana, tomato, coconut, soybean, wheat, balanca, rubber, winged bean, rice, oil palm, babassu, leucaena, cassava, neem tree, shee butter tree, and the bambara ground nut. A couple of additional titles in the series focus on more general topics such as seed banks and INBio, Costa Rica's National Biodiversity Institute (see Organizations).

Genetic Time Bomb
Date: 1994
Length: 50 minutes
Price: $150 (purchase); $45 (rental); low-income discount
Source: The Video Project

Erosion of genetic diversity in agricultural crops may result in severe food shortages in the near future. Thousands of traditional crop plants are becoming extinct as farmers replace them with high-yielding hybrid varieties. Genetic material that may help crops resist pests, drought, salinization, or other environmental challenges is being irretrievably lost. The video includes footage from around the world, as well as interviews with scientists, lawmakers, and "seed-savers" who cultivate rare "heirloom" varieties of vegetables and fruits.

Jaguar Trax
Date: 1996
Length: 34 minutes
Price: $65
Source: The Video Project

A high school science teacher spends a summer working on a banana plantation in Costa Rica but soon discovers that nearby farms are growing a variety of food crops more sustainably than the vast, chemical-dependent, monocropped banana plantations. The video includes interviews with Costa Rican elders, who share knowledge about medicinal plants and the importance of biodiversity.

Mayan Rainforest Farming
Date: 1987
Length: 29 minutes
Price: $49 (purchase); $20 (rental)
Source: Bullfrog Films

Mayan farmers have been farming sustainably for generations, using the multilayered rain forest as a model. In their traditional orchards, fruit trees shade vegetables and nonfood crops.

Plants in Peril

Date: 1988
Length: 26 minutes
Price: $89.95
Source: Films for the Humanities and Sciences

The current episode of mass extinctions differs from past crises in that for the first time in history, plants are disappearing as rapidly as animals. This video makes the case that whole ecosystems should be preserved in order to protect vulnerable plant species. Sites visited include a wetland area on the Atlantic seaboard where biologists are searching for the rare swamp pink; the Institute of Economic Botany at the New York Botanical Garden (see Organizations), where plants like the swamp pink are being conserved in *ex-situ* projects; and the University of California–Davis Germplasm Depository, which has amassed a large collection of South American wild tomatoes. Each tomato species possesses a different fundamental quality: one grows naturally in the desert and resists drought, another has extraordinary defenses against common pests, and some tolerate salty soils—a desirable quality since soils in many agricultural areas are salinizing. Researchers interbreed the varieties and come up with specialized varieties of tomatoes of great value to agriculture.

Seeds of Plenty, Seeds of Sorrow

Date: 1994
Length: 52 minutes
Price: $150 (purchase); $75 (rental)
Source: Bullfrog Films

In the 1960s and 1970s, the Green Revolution introduced new high-yielding but chemical-dependent crop varieties to traditional farmers in Third World countries. This video reveals how the Green Revolution led to a decline in India's traditional agricultural biodiversity, pesticide poisonings of workers, and severe socioeconomic degradation.

Restoration of Ecosystems

Making a Difference

Date: 1985
Length: 29 minutes
Price: $59.95
Source: The Video Project

Directed at middle and high school students, this three-part video profiles kids at work on behalf of the environment. One segment describes an effort to save wild mustangs from extinction, another shows young people planting trees in urban parks to mitigate air pollution, and the third is about teenagers working to bring salmon back to a stream near their home.

The Man Who Planted Trees

Date: 1987
Length: 30 minutes
Price: $95
Source: The Video Project

Nominated for an Academy Award, this animated video tells the story of a European shepherd who, through two world wars, planted and tended a beautiful forest on what had been an arid wasteland. This inspirational video shows that one individual's dedication is quite capable of changing the world.

The Return of Wolves to Yellowstone: Reclaiming a Habitat

Date: 1996
Length: 14 minutes
Price: $195
Source: New Dimension Media

This brief report traces society's demonization of wolves, the extermination of these canines earlier this century by ranchers in the western United States and Canada, and the decision by the U.S. Parks Service in the 1980s to reintroduce gray wolves to Yellowstone National Park. The controversial reintroduction program is illustrated with footage of the "wolf-lift" from Canada and examined through interviews with pro-wolf environmentalists and anti-wolf ranchers, accompanied by parks biologists' blow-by-blow accounts of the program's success.

Wild in the City: A Complete Guide to Growing Your Own Wildlife Oasis

Date: 1992
Length: 30 minutes
Price: $24.95
Source: Cambridge Science

Biodiversity cannot be limited to wild regions uninhabited by humans. This video advocates creating mini-refuges in urban areas and covers the steps involved: developing a habitat plan, choosing native plants that appeal to wild animals, creating the right conditions for them to thrive, and building a small artificial pond. A 48-page guide provides more specific directions.

Responses to the Biodiversity Crisis

Before It's Too Late
Date: 1993
Length: 48 minutes
Price: $79 (purchase); $40 (rental); low-income discount
Source: The Video Project

An examination of *ex-situ* conservation efforts to preserve the species in greatest danger of extinction, this video focuses on captive breeding programs undertaken at zoos and aquariums in New York, San Diego, Australia, England, and elsewhere. Successful reproduction by these species isn't enough to assure their long-term survival, however. In many cases there is no native habitat where they can be safely reintroduced.

Conserving America
Date: 1989
Length: Series of four programs, 58 minutes each
Price: $29.95 each or $99.95 for the series
Source: The Video Project

Produced in association with the National Wildlife Federation and originally broadcast on public television, this series documents conservation projects across the United States. "Champions of Wildlife" is about efforts to help individual endangered species by restoring their habitat and rehabilitating injured individuals. "The Rivers" examines efforts to preserve the 99 percent of American rivers that are not protected by the Federal Wild and Scenic Rivers Act. "The Challenge on the Coast" discusses the threats faced by coastal lands—pollution, erosion, and urban development—and how communities organize to help ease the problems. "The Wetlands," which profiles one of the world's most important yet least appreciated habitats, shows how activists have joined to preserve wetland refuges.

Crossing the Stones: A Portrait of Arne Naess
Date: 1993
Length: 47 minutes
Price: $250 (purchase); $75 (rental)
Source: Bullfrog Films

Arne Naess, a Norwegian philosopher, founded a movement called Deep Ecology, by which people acknowledge their deep ties to nature and commit themselves to environmental conservation. Naess is now in his eighties and lives with his wife in a secluded mountain cabin, but his enduring wonder at the natural world and his ability to distill confusing masses of information are still intact.

Flush Toilet, Goodbye
Date: 1987
Length: 29 minutes
Price: $195 (purchase); $50 (rental)
Source: Bullfrog Films

Physicist Pierre Lehmann designed a composting toilet to save money, energy, and potable water. The biodiversity of his backyard sewage treatment pond (its microbes and water plants) removes all impurities from the wastewater.

Greenbucks: The Challenge of Sustainable Development
Date: 1992
Length: 55 minutes
Price: $150 (purchase); $75 (rental); low-income discount
Source: The Video Project

Industry leaders from five continents show how their companies are using cleaner technologies, exploring alternative energy sources, wasting less in the production process, and producing recyclable products. Economists and futurists, including Alvin Toffler, attribute this change to more stringent regulations, better incentives, and the widespread realization that natural resources are limited.

GreenPlans
Date: 1995
Length: 56 minutes
Price: $95 (purchase); $45 (rental); low-income discount
Source: The Video Project

Green Plans are strategies formulated by government, industry, and private citizens for creating a sustainable economy in just one generation. The video visits two countries that have designed Green Plans, the Netherlands and New Zealand, and some communities in the United States that are trying to forge their own Green Plans. Featured also is Huey Johnson, the brain behind Green Plans. (See also the Print Resouces chapter for Johnson's book describing Green Plans.)

Heroes of the Earth

Date: 1993
Length: 45 minutes
Price: $59.95 (purchase); $35 (rental); low-income discount
Source: The Video Project

The Goldman Environmental Prize, considered the Nobel Prize of the environment, honors seven "eco-heroes" per year with its prestigious recognition (see Biographical Sketches). This video profiles each of the award recipients for 1993: a Colombian conservationist struggling to preserve the mountainous homeland of the Kogi people, a South Dakotan Native American woman fighting against toxic dumping on tribal lands, a Chinese woman who prevented a huge dam from being built on the Yangtse River, the leader of the former Soviet Union's first environmental movement, an Australian who fought to protect the largest sand island in the world, and two southern Africans who developed successful protection programs for rhinos and elephants.

How to Save the Earth

Date: 1993
Length: Series of six programs, 26 minutes each
Price: $175 each or $645 for the series (purchase); $45 each or $180 for the series (rental)
Source: Bullfrog Films

Each of the programs in this series focuses on two eco-heroes in different countries. Several of the heroes are working on biodiversity issues. "Taking the Waters" visits the former Czechoslovakia and Tunisia, where the heroes are organizing against dams that would destroy important wetlands. "The Monk, the Trees, and the Concrete Jungle" takes place in Thailand, site of vast rain forest destruction, and Japan, a major consumer of Thailand's wood. "How Much Is Enough?" addresses overpopulation in developing countries and overconsumption in superdeveloped ones—parallel threats to biodiversity.

Keeping the Earth

Date: 1996
Length: 27 minutes
Price: $29.95; low-income discount
Source: The Video Project

In this moving production, narrated by James Earl Jones, religious and scientific leaders of the United States join together to urge a new environmental ethic, one of dominion, not domination.

Natural Waste Water Treatment
Date: 1987
Length: 29 minutes
Price: $195 (purchase); $50 (rental)
Source: Bullfrog Films

Most large-scale sewage treatment plants are expensive, inefficient, and ineffective. The small-scale plants featured in this video rely on wetland flora to filter wastewater and result in cheap, efficient, and effective purification systems.

Our Future, Our Planet
Date: 1996
Length: 19 minutes
Price: $39.95
Source: The Video Project

This video covers the Youth Agenda portion of the 1992 Earth Summit, in which teenagers from all over the world met to define their role in the future of biodiversity conservation.

Profit the Earth
Date: 1990
Length: 58 minutes
Price: $24.95
Source: GPN

Six short segments each profile a "new environmentalist"—an activist who is helping resolve an environmental problem by compromising with rather than fighting against "the enemy." Zack Willey, for example, works on the water problem in southern California. For its water supply, Los Angeles diverts major rivers throughout the West, disrupting multiple aquatic ecosystems. At the same time, agribusiness is given water so cheaply that farmers have no incentive to conserve. Willey acts as a middleman, convincing farmers to implement conservation programs and sell their excess water to Los Angeles. Also included in the video are an Environmental Defense Fund economist who helped Congress hammer out an emissions trading program for power companies; an entrepreneur attempting to recycle disposable diapers; solar "farmers"; 3M's concerted attempts to reduce toxic emissions; and Hazel Henderson, a visionary economist who proposes adding the costs of environmental damage into the price of everyday products (i.e., a can of aerosol-propelled hairspray would cost the public $12,500). Each person is contributing in his or her small way, but taken together the work makes a big difference.

Race to Save the Planet
Date: 1991
Length: Series of 7 programs, 15 minutes each
Price: $89.95 each or $599 for the series
Source: Films for the Humanities and Sciences

Reviewing programs that seek to combat environmental problems such as solid waste disposal, overuse of pesticides, global warming, water pollution and scarcity, deforestation, and excessive waste of energy, the short segments in this series are designed to stimulate students to analyze environmental policy and human impact on the natural world.

Rain Forests: Proving Their Worth
Date: 1990
Length: 30 minutes
Price: $85 (purchase); $45 (rental); low-income discount
Source: The Video Project

One approach to saving the rain forest has been to harvest certain valuable products from rain forests and market them abroad to environmentally conscious consumers. Cosmetics, foods, and handicrafts harvested from tropical rain forests are already being sold throughout the world. This program explores the conservation successes of this approach and the obstacles remaining.

Spirit and Nature
Date: 1991
Length: 88 minutes
Price: $35
Source: The Video Project

Biodiversity and its conservation cannot be left to scientists. This program is composed of conversations between thoughtful host Bill Moyers and spokespeople from five world religious traditions. They share spiritual and ethical responses to the global environmental crisis and insights from their own cultures on how people everywhere can develop a more equitable relationship with nature.

Turning the Tide
Date: 1988
Length: Series of 7 programs, 26 minutes each
Price: $975
Source: Bullfrog Films

The programs in the series, hosted by British scientist David Bellamy, cover major environmental challenges to the planet, including overconsumption of natural resources, overpopulation, contamination of drinking water, the waste and destruction caused by huge energy-producing projects like dams, and the erosion of genetic diversity. Bellamy is convinced that we have the tools to reverse the destruction but that our commitment to do so is still lacking.

Voices of the Land
Date: 1991
Length: 21 minutes
Price: $195 (purchase); $40 (rental)
Source: Bullfrog Films

Preserving biodiversity will require a deeper spiritual connection between people and nature. Three articulate voices explain why people traditionally consider certain places sacred and how important it is to let nature nurture the soul. Interviewees include a Southern Ute elder, a native Hawaiian, and Dave Foreman, cofounder of Earth First! (see Organizations).

Worth Quoting with Hazel Henderson
Date: 1994
Length: Series of 9 programs, 30 minutes each (on 4 cassettes)
Price: $59–$89 per cassette or $195 for the series (purchase); $35–$45 per cassette (rental)
Source: Bullfrog Films

Hazel Henderson is known for her economic analyses incorporating the true environmental cost of products and services into their price. This series of conversations between Henderson and other pioneers of alternative thought includes segments entitled "Inner Roots of Our Global Crises," "Planetary Ecology vs. World Trade," and "Agenda 21: North and South Views of the 21st Century."

Audiotapes

The following audiotapes are of interviews or speeches by scholars and political analysts on various biodiversity/environmental themes. They are also available in transcript form. All are produced by Alternative Radio.

Biopiracy
Speaker: Vandana Shiva
Date: 1997

Vandana Shiva reveals the dangers that globalization presents to the conservation of biodiversity in the agricultural sector and the economic motives behind biotechnology.

Environmental Racism
Speaker: Robert Bullard
Date: 1992

Bullard describes how minority communities are frequently targeted for toxic waste dumps and as sites for toxic industries.

Environmental Racism in Hawaii and the Pacific Basin
Speaker: Haunani-Kay Trask
Date: 1993

Trask gives an indigenous Hawaiian view of the environmental consequences of the conquest of the Pacific and the militarization of Hawaii.

Ethical Anaesthesia
Interviewee: Vandana Shiva
Date: 1997

Indian activist Shiva discusses some of the biodiversity conservation issues she is most concerned about, including genetic erosion of traditional crops and multinational industry's new practice of patenting genes.

The Fate of the Amazon Rainforest
Speaker: Alexander Cockburn
Date: 1989

A speech outlining the causes of rain forest destruction described in Cockburn's book, *The Fate of the Amazon*.

The Green Belt Movement of Kenya
Speaker: Wangari Matthai
Date: 1990

The founder of this strong reforestation movement in Kenya describes her work and the challenges she faces.

An Indigenous View of North America
Speaker: Winona LaDuke
Date: 1993

The Native American reverence and respect for the natural environment is completely at odds with the Euro-American view of nature as a wild force that must be harnessed to produce income.

The Media and the Environment
Speaker: Alexander Cockburn
Date: 1992

Media analyst Cockburn comments on how mass media has negatively stereotyped the environmental movement.

Plundering Paradise: The Struggle for the Environment in the Philippines
Interviewee: Robin Broad
Date: 1993

Robin Broad describes the environmental challenges facing Philippine environmentalists: massive destruction of mangroves by the aquaculture industry and extensive rain forest deforestation.

Rain Forest Politics
Interviewee: Alexander Cockburn
Date: 1989

The author of *The Fate of the Amazon* discusses the political causes of rain forest destruction in Brazil.

Recovering the Commons
Speaker: Vandana Shiva
Date: 1994

The commons in twelfth- and thirteenth-century Britain were public lands worked by villagers in collaboration. Later in history, they were privatized, divided up, and sold to wealthy residents. Shiva discusses her vision for reinstating the commons tradition.

Technology, Spirituality, and the Future of the Planet
Speaker: Helen Caldicott
Date: 1995

A passionate, effective speaker, Caldicott describes the sustainable future possible once technology and spirituality are allowed to intersect.

Visions of the Environmental Movement
Speaker: David Brower
Date: 1993

The U.S. environmental movement's revered "grandfather" looks to the future and describes the direction he thinks environmentalists should be headed.

World Bank/IMF: Fifty Years Is Enough
Speaker: Danny Kennedy
Date: 1994

The World Bank has spent 50 years funding huge economic development projects, some of which have had disastrous environmental effects.

Contact Information for Distributors and Vendors

Alternative Radio
P.O. Box 551
Boulder, CO 80306
(800) 444-1977
Web site: http://www.freespeech.org/alternativeradio

BBC
P.O. Box 2284
South Burlington, VT 05407-2284
(800) 435-5685
Fax: (802) 864-9846
Web site: http://www.bbc-worldwide-americas.com

BFA Educational Media
468 Park Avenue South
New York, NY 10016
(800) 221-1274
Fax: (314) 569-2834

BIOSIS
2100 Arch Street
Philadelphia, PA 19103-1399
(800) 523-4806
Fax: (215) 587-2016
Web site: http://www.biosis.org

Bullfrog Films
P.O. Box 149
Oley, PA 19547

(800) 543-3764
Fax: (610) 370-1978
E-mail: bullfrog@igc.apc.org

Cambridge Science
P.O. Box 2153, Dept. SC4
Charleston, WV 25328-2153
(800) 468-4227
Fax: (800) 329-6687
Web site: http://www.cambridgeol.com/cambridge/

Cambridge Scientific Abstracts
7200 Wisconsin Avenue, #601
Bethesda, MD 20814
(800) 843-7751
Fax: (301) 961-6720

The Cinema Guild
(800) 723-5522
Fax: (212) 246-5525
Web site: http://www.cinemaguild.com

Congressional Information Service, Inc.
4520 East-West Highway, #800
Bethesda, MD 20814-3389
(800) 638-8380
Fax: (301) 654-4033
Web site: http://www.cispubs.com

EBSCO
10 Estes Street
Ipswich, MA 01938
(508) 356-6500
Fax: (508) 356-9372
Web site: http://www.epnet.com

Environmental Media
P.O. Box 99
Beaufort, SC 29901-0099
(800) 368-3382
Fax: (803) 986-9093
Web site: http://www.envmedia.com

Filmmakers Library
124 East 40th Street
New York, NY 10016
(212) 808-4980
Fax: (212) 808-4983

Films for the Humanities and Sciences
P.O. Box 2053
Princeton, NJ 08543-2053
(800) 257-5126
Fax: (609) 275-3767
E-mail: custserv@films.com

Glencoe/McGraw-Hill
Order Department
P.O. Box 543
Blacklick, OH 43004-0543
(800) 334-7344
Fax: (614) 860-1877
Web site: http://www.glencoe.com

Great Plains National (GPN)
P.O. Box 80669
Lincoln, NE 68501-0669
(800) 228-4630
Fax: (800) 306-2330
E-mail: gpn@unl.edu

Hawkhill Video
125 East Gilman Street
P.O. Box 1029
Madison, WI 53701-1029
(800) 422-4295
Fax: (608) 251-3924
Web site: http://www.hawkhill.com

Home Vision
5547 North Ravenswood Avenue
Chicago, IL 60640-1199
(800) 826-3456
Fax: (773) 878-8406
E-mail: classics@homevision.publicmedia.com

Information Access Company
362 Lakeside Drive
Foster City, CA 94404
(800) 227-8431
Fax: (800) 700-1890
Web site: http://library.iacnet.com

Institute for Global Communications
Presidio Building, #1012, First Floor
Troney Avenue
P.O. Box 29904
San Francisco, CA 94129-0904
Web site: http://www.igc.org

Knight Ridder Information, Inc.
P.O. Box 10010
Palo Alto, CA 94303-9620
(800) 334-2564
Fax: (415) 254-8132
Web site: http://www.krinfo.com

Landmark Media, Inc.
3450 Slade Run Drive
Falls Church, VA 22042
(800) 342-4336
Fax: (703) 536-9540
E-mail: Landmrkmed@aol.com

National Geographic Society
Educational Services
1145 17th Street NW
Washington, DC 20036-4688
(800) 368-2728
Fax: (301) 921-1575
Web site: http://www.nationalgeographic.com

National Information Center for Educational Media
Access Innovations
P.O. Box 40130
Albuquerque, NM 86196
(800) 926-8328
Fax: (505) 256-1080
Web site: http://www.nicem.com

National Information Service Corporation
Wyman Towers, Suite 6
3100 St. Paul Street
Baltimore, MD 21218
(410) 243-0797
Fax: (410) 243-0982
Web site: http://www.nisc.com

New Castle Communications, Inc.
229 King Street
Chappaqua, NY 10514
(800) 723-1263
Fax: (914) 238-8445
E-mail: ideas@newcastlecom.com

New Dimension Media, Inc.
611 East State
Jacksonville, IL 62650
(800) 288-4456
Fax: (800) 242-2288
E-mail: btsb@fgi.net

Newsbank Reference Service
58 Pine Street
New Canaan, CT 06840-5426
(800) 243-7694
Fax: (941) 263-3004
Web site: http://www.newsbank.com

Online Computer Library Center (OCLC)
6565 Frantz Road
Dublin, OH 43017
(800) 848-5800
Web site: http://www.oclc.org

Optilearn, Inc.
15 Park Ridge Drive, #200
P.O. Box 997
Stevens Point, WI 54481
(800) 850-9480
Fax: (715) 344-1066

PBS Video
1320 Braddock Place
Alexandria, VA 22314-1698
(800) 344-3337
Fax: (703) 739-5269
Web site: http://www.pbs.org

Public Media Education
5547 North Ravenswood Avenue
Chicago, IL 60640-1199
(800) 343-4312
Fax: (773) 878-0416
E-mail: classics@homevision.publicmedia.com

Social Issues Resources Series (SIRS)
P.O. Box 2348
Boca Raton, FL 33427-4704
(800) 232-7477
Fax: (561) 994-4704
Web site: http://www.sirs.com

SVE & Churchill Media
6677 North Northwest Highway
Chicago, IL 60631
(800) 829-1900
Fax: (800) 624-1678
Web site: http://www.svemedia.com

Turner MultiMedia
One CNN Center
P.O. Box 105780
Atlanta, GA 30348-5780
(800) 639-7797
Fax: (404) 827-1775
Web site: http://www.learning.turner.com

TV Ontario
U.S. Sales Office
1140 Kildaire Farm Road, #308
Cary, NC 27511
(800) 331-9566
Fax: (919) 380-0961
E-mail: U.S.sales@tvo.org
Web site: http://www.tvo.org/sales

The Video Project
200 Estates Drove
Ben Lomond, CA 95005
(800) 475-2638
Fax: (408) 336-2168
E-mail: videoproject@igc.apc.org

World Wildlife Fund
Department CA4
P.O. Box 4866
Hampden Post Office
Baltimore, MD 21211
(410) 516-6951
Fax: (410) 516-6998

Glossary

adaptive radiation The type of evolution that occurs when a single species branches out and forms many species, each adapted to fill different ecological niches. A famous example of this type of evolution is the 14 species of Darwin's finches on the Galápagos Islands, all descended from a common ancestor.

arthropod A phylum in the kingdom of animals that includes such insects and crustaceans as centipedes, millipedes, insects, crabs, lobsters, shrimp, spiders, scorpions, mites, ticks, etc. Scientists presume that this phylum includes more species than any other.

biome This is the largest type of ecological region recognized by biologists. Biomes refer to the type of habitat rather than a certain habitat itself. For example, the tropical rain forest is a biome, whereas the Amazon Basin rain forest is a specific ecosystem.

biophilia A word coined by biologist E. O. Wilson. It refers to the innate fascination that humans have with all life-forms.

bioprospecting The practice of searching for useful compounds occurring naturally in ecological communities.

biosphere The thin skin of life covering the earth's surface, including soil and the microorganisms inhabiting it, vegetation, animals, and all other living beings inhabiting the planet.

biotechnology Any technique used to make or modify a living organism. A simple, age-old example is the use of yeast to produce bread. More recent examples include the use of certain bacteria to "eat up" toxic waste and the alteration of sheep genes to produce super milk producers.

carbon dioxide sinks Ecosystems such as forests or communities of oceanic plankton that absorb large amounts of carbon dioxide and produce a great quantity of oxygen.

CITES The Convention on International Trade in Endangered Species, an international treaty regulating trade in endangered species, adopted in 1973.

conservation biology A new multidisciplinary field linking biology with other fields that contribute to conservation, such as political science, economics, and philosophy.

debt-for-nature swap A method of financing conservation in biodiverse countries with a large foreign debt. An international conservation organization purchases a portion of the country's debt and trades it back to the country in return for the country's commitment to conserve a certain ecologically important area.

ecological community All of the organisms that coexist in a particular habitat, an area ranging in size from a tiny niche to an ecosystem to a landscape to a biome to the biosphere itself.

ecosystem A community distinguishable by certain characteristic environmental conditions such as elevation, temperature, moisture level, soil type, wind, and amount of sunlight; the plants and animals that live there; and the interactions among them.

ecotourist An ecotourist is interested in the natural history of his/her destination, that is, its ecology and the local culture. This type of tourist helps fund conservation projects because ecotourists pay entrance fees to the natural reserves they visit. If a local economy is dependent upon ecotourism, local people will in theory be more motivated to conserve the natural areas ecotourists want to visit.

Endangered Species Act Enacted in 1973, this legislation requires the U.S. Fish and Wildlife Service and the National Marine Fisheries Service to list species that are endangered (in danger of extinction in its entire range) or threatened (likely to become endangered soon) and to work for their recoveries. It also prohibits hunting these species and destroying their habitats. Also known as the ESA.

endemism A species is endemic (the adjectival form of endemism) to a certain location if it originated there and is confined to that place. Darwin's finches, for example, are endemic to the Galápagos Islands.

entomology The branch of zoology (the study of animals) that focuses on insects. A practitioner in this field is called an entomologist.

epiphyte Epiphytes are plants that attach to a host instead of rooting in the ground. Bromeliads and orchids, both forms of epiphytes, attach to tree limbs in order to receive more sunlight than they would get on a forest floor and absorb moisture as it drips down the tree trunk. They differ from parasites in that they do not harm their host.

ESA Endangered Species Act. See above.

ethnobotany This two-part word refers to the study of people's (*ethno* means a race or cultural group) use of plants *(botany)*. Ethnobotanists study the plants that different peoples use as medicine, food, hallucinogens, etc.

eutrophication A lake or pond is said to be eutrophic if it grows excessive algae and becomes oxygen-depleted. Eutrophication occurs in shallow bodies of water where the penetration of sunlight and abundant nutrients (such as dissolved phosphates from fertilizer) combine to provide perfect conditions for algae blooms. Because the algae consume all the available oxygen, they effectively suffocate the lake and make it an inhospitable habitat for any other form of life.

evolution Changes occurring within a species over generations that result in descendants differing in form and function from their ancestors. Through the process of *natural selection* (see below), evolution gradually makes a species better adapted to its environment. *Vertical evolution* is the gradual change of a species until it is so different from the original version that it becomes a new species. See also *adaptive radiation*.

***ex-situ* conservation** *Ex* (outside)-*situ* (place) refers to conservation of species outside of their natural habitat, that is, in captive breeding centers, laboratories, and botanical gardens. *Ex-situ* approaches to conservation are usually undertaken as a last resort, when scientists judge that the species will become extinct without intervention. Compare to *in-situ* conservation.

FAO Food and Agriculture Organization, an agency of the United Nations.

Gaia Hypothesis Proposed by Drs. James Lovelock and Lynn Margulis, the Gaia Hypothesis posits that the biosphere is a self-regulating mechanism. By varying its production of carbon dioxide, oxygen, and other chemical elements, the biosphere maintains the atmospheric conditions that perpetuate life.

gene The unit on a chromosome that is hereditary and manifests as a specific trait of the organism in which it occurs.

Green Revolution Promoted by development workers in the 1960s in poor, agricultural countries, the Green Revolution encouraged farmers to switch from traditional methods and crops to lab-engineered varieties that were highly productive but depended upon expensive inputs such as fertilizers and pesticides. The Green Revolution is now seen by many as responsible for genetic erosion in agriculture.

habitat A type of environment (such as a tropical rain forest) or a particular environment itself (the rain forest at La Selva, Costa Rica). Habitat is defined for a particular species. For example, a jaguar's habitat measures thousands of square kilometers because it has a very large range. A bacterium's habitat may be microscopically small.

heirloom varieties An agricultural or horticultural term referring to unique varieties of plants developed and conserved by a family or a community. Also referred to as land races, folk varieties, or crop ecotypes.

host-specific This adjective refers to species that feed on only one other species. Many hummingbirds, for example, are able to gain nourishment from only one of the hundreds of flowering plants in a tropical rain forest. They would be considered host-specific.

indicator species Species whose disappearance could signal the ecological breakdown of a habitat.

indigenous This adjective describes peoples who are descendants of the precolonial inhabitants of a country. Their culture and lifestyle are usually distinct from the descendants of colonists. Indigenous peoples make up about 4 percent of the world's population.

***in-situ* conservation** This approach to conservation refers to conservation of a natural habitat. Compare to *ex-situ* conservation.

intertidal An adjective describing a coastal area lying between the low and high tide points.

IUCN World Conservation Union (see the Organizations chapter).

keystone species Species that play a crucial role in maintaining the overall health of an ecosystem. They may stand as the base of a food web, provide habitat for species, or control overpopulation of their prey. Often their identities are revealed only after they disappear from an ecosystem.

landscape A conglomeration of adjacent ecosystems in a defined region such as a watershed.

marsupial The order of mammals whose members spend the later part of their gestation in a pouch (marsupium) on their mother's body.

monocrop Today's large-scale farmers usually practice monocropping, or the cultivation of a single crop per field. Though this farming method simplifies mechanized harvest, it decreases the diversity of organisms coexisting in the same area.

natural selection One of the processes by which evolution occurs. Natural selection refers to the principle that individuals with favorable traits will reproduce more successfully than others and, as a result, their genes will eventually dominate the gene pool of the species or population.

niche A space, either physical or in the organizational structure of an ecological community, that a species adapts to fill. For example, night-blooming flowers in a tropical rain forest provide food material for nocturnal pollinators. Ecosystems with a wide range of niches are more likely to have a high number of species.

organism Any living being, plant or animal, fungus, protozoan, or bacterium.

ozone shield The layer of ozone—a form of oxygen—in the upper atmosphere that screens out intense rays of the sun.

photosynthesis The process by which plants convert solar energy into sugars that fuel their growth. During the process of photosynthesis, plants absorb carbon dioxide and generate oxygen.

placental mammals Mammals that gestate within their mother's uterus and are nourished by a placenta.

population A population of a species is a group of individuals living in a geographically separate area from other members of the same species.

speciation The evolutionary process by which distinct species branch off from their common ancestor.

species The smallest taxonomic category of living organisms generally accepted by biologists. The main defining characteristic of a species is that its members interbreed but will not mate with others of different species.

sustainable development The use of natural resources by a group of people at a rate gradual enough that those resources are allowed to replenish.

taxa Taxonomic categories, which include kingdom, phylum, class, order, family, genus, and species. Also called taxonomic divisions.

taxonomy The science of classifying forms of life. The physical features of an organism and its evolutionary history are both important considerations in its classification.

UNEP United Nations Environment Programme (see the Organizations chapter).

UNESCO United Nations Educational, Scientific and Cultural Organization.

watershed The region that drains into a river, a river system, or a body of water.

WWF World Wildlife Fund (see the Organizations chapter).

Index

Page numbers in bold print denote main entry.

Anne Becher began work on this book while living in what could be considered a world capital of biodiversity, Costa Rica. She coauthors a guide for responsible eco-travel in that country, *The New Key to Costa Rica* (Berkeley, CA: Ulysses Press). Becher holds a masters degree in Hispanic Linguistics (University of Colorado, 1992) and was a Watson Fellow in 1987–1988, traveling to Argentina to study the Mothers of the Disappeared. She and her husband backpacked up the spine of the Americas, passing through most of the countries in South and Central America. Her many articles, on topics ranging from prenatal fitness to religious freedom in Sandinista Nicaragua, have been published in a variety of periodicals throughout the hemisphere. With her husband and their two children, Becher now lives in Boulder, Colorado, where she teaches Spanish at the University of Colorado.